Acknowledgements

We would like to thank colleagues who reviewed a draft of this manuscript and offered meaningful suggestions for improvement, including Allison Beardsley, Anneliese Johnson, Dana Smith, Kris Bakkum, Mary Christopherson, Tanya Haugen, and Teresa Shume. We also thank the many teachers, students, colleagues, and friends who have shared their thoughts and insights about science, teaching, and learning over the years.

Ewww...

There's Some *Geology* in my Fiction!

An anthology of YA earth science short stories with ideas for classroom investigation

Russ Colson and Mary Colson (eds.)

With contributions from Mary Alexandra Agner, Robert Dawson, Douglas Smith, Jacqueline Howland, S. Q. Eries, Mary Ibe, L. L. Hill, Justin Short, Kate Kelly, and Scott Schad

A Real Dragon Press Book

Ewww, There's some Geology in my Fiction
A Real Dragon Press book
12158 230th St N
Hawley MN 56549

https://realdragonpress.com

Cover Art by Erin Colson
Cover Design by Russ Colson
Editorial work on stories by Laurie Sand

Statement of AI: No generative AI was used in creating this book.

Printed Version ISBN 978-0-9986321-7-9

2026, Real Dragon Press

Table of Contents

Environment/Climate

Permafrost by Kate Kelly

Focus of the example investigation: modeling, mathematical thinking

Feedstock by Scott Schad

Focus of the example investigation: discussion, experimentation, interpretation

Introduction

Ewww, There's Some Geology in My Fiction offers teachers and students a collection of earth-and-space-science-related short stories appropriate for a middle or high school classroom along with accompanying prompts for developing an investigation and with example investigations that we have tried. We, the editors Russ and Mary, hope that the book provides an opportunity to practice investigative thinking in a non-threatening environment where the goal is to ask a question and pursue an answer without the constraint of getting the "textbook" answer.

This book is not a curriculum or collection of plug-and-play activities, and so the science teacher might reasonably ask, what, then, is it good for? The answer to this question requires some background context and discussion.

Learning answers is not at all the same as doing science, whether we are a teacher, student, or scientist. Creating an investigative thinking experience in the classroom—that is, an engagement with the full sweep of the practice of doing science—is much more involved, harder to teach, and more challenging to assess than offering information or explaining and applying theories. Investigative thinking requires that we all, as fellow investigators, 1) ask our own questions, 2) figure out how to address those questions with observation, experiment, or analogy, 3) interpret the meaning and implications of what we see and measure, 4) develop a theoretical model to explain our observations, and 5) reason through the evidence to come up with convincing arguments that connect observational evidence to explanatory models.

Unless we already know the outcome ahead of time, or are following someone else's preset recipe for investigation, each of these steps requires creative problem-solving and a spontaneous pursuit of pathways that we did not know to follow. Thus, *investigative thinking* is not a series of well-defined steps, like the old "scientific method," but rather it is a fluid *practice*, a way of approaching the understanding of science that does not depend on trusting the "experts" or searching online for "answers." Authentic investigation engages us in the centuries-old process of figuring out how our world works.

If you, as teacher, student, or scientist, buy-into the premise above—that real investigation involves figuring out something that you don't already know and not merely following a fixed set of procedural steps to a predestined ending—then it follows that **true investigative thinking is *intrinsically incompatible* with any fixed curriculum or set of instructional standards** (yes, we the editors know that many educators are arguing otherwise, but this is our view based on a lifetime of experience with investigative teaching).

Authentic investigation involves both creative thinking and logical deduction outside the bounds of any procedural systems developed and proposed for the classroom. It goes beyond any curriculum that purports to assemble "what every student should know."

We, the editors, propose that the stories and activities in this book provide an opportunity for students to try out their investigative thinking in a self-contained way that is unconnected to the expectations of their textbook. We hope this frees the student to experience asking questions and pursuing answers through investigation, which involves much more than just looking stuff up.

We also hope that this book can provide a similar experience for the teacher.

Why is that necessary?

Curricula and procedural systems are the norm in our classrooms because following preset steps and learning the information is comfortably achievable for both teacher and student. We, both teachers and students, know what we're supposed to do and how to do it.

But what if, like scientists through the ages, we are only presented with a universe full of questions and no predetermined knowledge or set of procedural steps to address them?

Woah! How can I, as a teacher with limited research experience in my discipline, authentically investigate, or mentor students through an investigation, if I don't know what's supposed to happen, if I can't just follow a curriculum, or a set of steps, or an experimental recipe? Where do I even start?

First, practice the practice! Doing or mentoring investigations requires confidence in one's ability to investigate, which, in turn, requires practiced experience at real research or problem solving. Confidence and experience support a teacher in creating investigative steps in a situation where a predetermined set of steps is unlikely to bear fruit.

Leaping into the unknown and figuring things out requires its own set of skills, different from following procedural steps or knowing information. These skills can only be gained through **practice and experience** at "trying it out." Scientists typically gain this experience through their own extensive research projects as graduate students. Teachers must often gain this experience through diving into investigation within their own classrooms. This book is intended to offer opportunities to gain a type of research-like experience, supplemental to or preparatory to engaging in your own classroom investigations

We, the editors, propose that the stories and activities in this book provide an opportunity to try out your own thinking and investigative skills until you gain confidence in your ability to dive into the unknown and come out with something meaningful.

We teachers need experience with many investigations of our own in order to mentor our students through the sometimes-confusing process of investigation. To get that experience, we, like our students, need to struggle through the fog of doing real investigation that does not come from a textbook. All of the classroom management skills in the world can't help us engage our students in investigative thinking if we ourselves don't understand what that means at a visceral and experiential level. In this book, we are focused on gaining experience with investigative thinking, not on strategies to manage classroom investigations.

With this goal, this book offers three avenues toward textbook-free investigation: 1) a small sandbox to play in, 2) opportunities to chase our own rabbits, and 3) a chance to learn investigation by example.

A Small Sandbox in which to Play

The world is big and complex, and asking questions of the world can feel intimidating even for experts. For example, "what causes climate change" or "how do I know climate change is happening" are mindbogglingly complex questions that cannot be readily addressed through investigation in the classroom and sometimes might even encourage either broad theoretical speculation or politically-motivated assertions.

Short stories can provide a fun way to stir investigative thinking, constrain the options of what to investigate, and allow for a wide range of questions and investigative procedures.

A fictional story with an earth science element offers a contained and engaging environment in which to ask and answer questions. With a prompt like a story, we and our students may not experience the same crippling need to get the "right answer." Instead, we can be driven by our own curiosity.

The stories in this book offer a non-threatening way for us, as students and teachers, to engage in our own investigative thinking within the confines of a "sandbox" that both stirs questions and constrains those questions to a reasonably small subset of possibilities[1].

To summarize, both student and teacher need practice at small investigations before a larger investigation becomes feasible to tackle. This little book provides small opportunities to practice the practices of science at a non-threatening scale, giving experience on which to base creative and spontaneous steps in the future. The investigations born from these readings can be as complex or simple as the reader is ready to pursue, allowing each person to "chase their own rabbits" at whatever level they are ready for.

Chase Your Own Rabbits

When I (Mary) was a student in geology, I asked one of my professors, Bob Schwartz, "Why did you decide to get a Ph.D.?"

"I wanted to chase my own rabbits," he said.

I was visiting Bob and his wife, Marie, a fellow geology major at Allegheny College in Pennsylvania, at their farm in northwest Pennsylvania. We were sitting on the porch steps talking and watching the sun set. I was still casting about for who I was going to be in life and what I was going to do, and the question arose from a deep personal interest.

I have remembered his unexpected response for over forty years. My take-away was not that one has to get a Ph.D. to chase their own rabbits, but rather that here was someone so excited by what he didn't know, curious about what he wanted to be curious about, and hopeful to

---

[1] We discuss the importance of providing guiding prompts and limiting classroom options in our book *Learning to Read the Earth and Sky* Colson R. O. and Colson M C. (2016) Explorations Supporting the NGSS, Grades 6-12, NSTA Press, 426p.

discover, that he chose a challenging path in life where he could pursue that excitement, curiosity, and hope. Not only did Bob have no fear of "not knowing the answer," but not knowing the answer formed the very basis for the playful curiosity that shaped his life.

Students and teachers can be attracted by that playful curiosity—finding fun, and perhaps a deeper understanding of what science really is all about—by chasing their own rabbits!

The stories in this book offer opportunities for us as students and teachers to practice chasing our own rabbits—asking a question stirred by the story, developing a plan for how to answer that question that does not involve Google and various AI engines, and then implementing that plan and interpreting our results.

Questions can be broad or narrow, although narrower questions have a better chance of being addressed in a reasonable amount of time. Questions can be varied: Does what happens in the story make scientific sense? How did the heroes in the story figure out their puzzle and would that work in real life? What scientific methodology or relationship is implied, but perhaps not discussed, in the story that is necessary to make the outcome of the story reasonable or possible? Is the premise of the story true? If there is an implied mathematical relationship in the story, such as the amount of energy needed or the mass of carbon sequestered, then do the numbers add up and make sense in a realistic world?

The plan and investigation can likewise be varied, and as complex or simple as we choose. We can develop ways to pursue questions with experimental investigations, analogous puzzles, modeling exercises, mathematical thinking, or informed discussions.

All of these pursuits offer opportunities to "try out" our investigative skills and learn to investigate from that experience.

We all recognize, as students and teachers, that it is easier to ask deeper questions and develop our own investigative procedures when we have a greater body of knowledge about earth science to start with, but stories allow us to start where we are and move forward from there. The contained nature of both the stories and exercises offered here provides flexibility to pursue one's own rabbits as far as time, prior knowledge, and interest allow. We can push the boundaries of our own experience with the practice of science and become more able to tackle the next authentic investigation to come our way.

One thing to keep in mind as we chase our own rabbits: investigation needs to go beyond asking an engaging question and then formulating a vague notion of how to pursue the question. It's great to ask questions and come up with general steps of how we might pursue answering them, but investigative thinking requires developing a detailed process for pursuing that question, going through that process, and revising it when it doesn't pan out like we expected. That process will depend on our particular question and situation and will almost never be a generic set of steps patterned after some other investigation. If we wonder how characters in the story figured out which dinosaur bone is older, how can we model the rock layers to understand? How can we figure it out *in detail*? If we wonder if the energy from a tsunami could be countered by energy from windmills, what calculations are needed to test that? Can we work through those calculations? If we wonder whether ice would really get stronger if it got colder, how can we design an experiment to test it? Can we implement the experiment and see if it actually gives us an answer?

Unless we plan to spend the next month or year (or lifetime) on a particular question, we will probably need to limit our question to some small part of a full investigation, but it is better to examine that small part with great rigor than to examine a much larger part with vagueness. What's more, we need to actually *go through* our process of investigation, not just think about it abstractly. It is in the process of *doing* that we learn what investigation is all about. Investigative thinking must be iterative, revising our plan on the fly as we learn new things about our question.

Just like no military plan survives the first contact with the enemy, so no investigative process survives the first results of actually doing it. It can be frustrating to develop investigative steps only to see them not quite work out. However, gaining experience from many such investigations can inform our next one, and allow us to more quickly recognize what is not working and develop an alternative plan. **It is in this iterative process—trying and revising—that authentic investigation occurs and we gain experience that allows us to do the next investigation better and to mentor students through the sometimes-confusing fog of investigation.**

Learn by Example

Doing our own investigating is an essential element of authentic investigation, including asking our own questions and figuring out our own procedural steps. However**, seeing someone else's question, and the steps that they developed to pursue it, can also give us experience with investigation and inform our next effort at investigative creativity.**

As Isaac Newton famously said, "we stand on the shoulders of giants." As we, the editors, have emphasized in this book and in *Learning to Read the Earth and Sky*, a teacher in the classroom should not be only a curriculum manager, but must be a mentor, actively engaged with students in such a way as to both show their own passion for learning and illustrate through example how investigation can take place. We teachers are among the giants on whose shoulders our students must stand. They learn how to investigate from watching our *example*, not simply from hearing our *instruction*.

Given this philosophy, it's reasonable to think that we, the editors of this book, should offer some of our own "mentoring." Consequently, for each story, we offer examples of our own investigative thinking. These examples should not substitute for the need for all of us, teachers and students alike, to create our own rabbit-chasing investigations, but they can offer insight into how we, the editors, approach asking questions and pursuing answers for each of these stories, perhaps spurring wider ideas or offering further experience in what investigation entails.

We, the editors, have tried to offer stories and example investigations that address a breadth of the earth sciences (including some geology, astronomy, chemistry, engineering, and environmental science) and also a variety of the different practices of science (including experimentation, interpretation, modeling, mathematical thinking, and discussion). One way to categorize this effort is given in the Table of Contents above, which suggests the main science field that might be associated with each story and also indicates the practices of science that we chose to focus on in our example investigations. This outline might provide insight into how teachers could use different stories in their own curriculum, although most stories could reasonably fit into a variety of curricular spots.

Some key ideas that emerge from our example investigations might be the following:

- In this book, we use the word "investigation" to encompass the entire scope of the doing of science. However, not all of the NGSS Science and Engineering Practices need to be included for an investigation to be investigative. Pieces of investigation are also investigation, and, in fact, most real-world research involves scientists doing small parts of much larger investigative efforts. It is often more achievable in the classroom to target particular practices rather than attempting to do all of them at once.
- A corollary to the above assertion is that "investigation" does not automatically mean "experiment," although we argue that observational evidence must always be at the heart of investigation in some way for it to be considered a scientific investigation.

- Inventing an experiment is an exercise in engineering design. The experimenter must figure out what materials to use, how to put those materials together, how to optimize results, and how to revise it all when the first effort doesn't work.
- Scientists don't have an answer key. In like fashion, we all, scientist, teacher, and student, have to figure out from our data and observations whether our results are reliable, whether they make sense or not, and what the results mean. There isn't an "outside source" that tells us that we got it correct.
- None of us can break the laws of nature. If we get a particular result, it means something real (even if what is "real" is only that we read the ruler wrong). Figuring out ALL of the results that we *actually got* (not what the textbook said we should get), and why those results make sense, is what data interpretation is about. "We got the wrong answer because of experimenter error" is a cop-out, not an analysis or interpretation of data.
- Not all questions can be investigated scientifically. Part of investigative thinking is learning to recognize what can be investigated through scientific practices and what cannot.
- The key to learning to do science is to jump in and try to do it. Learn by doing. We also learn by seeing what others have done. This is the foundation for the belief that teachers should be mentors in the classroom, participating in investigation rather than simply overseeing it. The need to learn from others is also the reason for the example investigations in this book.

Final Thoughts

These stories and lessons are not intended to be a curriculum or provide complete "pre-canned" investigative activities. So where might these stories and investigations fit into a curriculum?

Maybe as a fun way to kick off a new unit on astronomy, stories in rocks, climate change, or plate tectonics. Maybe as a capstone activity at the end of a unit, a reward for a completed effort. Maybe as a refreshing break in the middle of a unit. Any of these opportunities might add a bit more authentic free-form investigative thinking into the unit, a chance for teachers and students to pursue their own rabbits, to ask questions, to figure out investigative steps, and to draw their own conclusions without consuming an entire unit in a more expansive investigation (which we also encourage where possible!). In short, these stories and prompts can offer a way to learn what real science is all about!

<u>A note about readability and grade-level appropriateness</u>: The short stories offered in this book are generally appropriate for middle school and up. We have used the admittedly-imperfect but widely-available Flesch-Kincaid readability tests to evaluate reading ease and grade-level. The Flesch-Kincaid ease of reading scores for the fictional short stories range from a low of 68.1 for "Feedstock," to a high of 87.5 for "Permafrost," with most others falling in the mid- to-upper 70s. The higher Flesch-Kincaid numbers indicate greater ease of reading, and scores in the range reported here correspond generally to a seventh-grade reading level. The Flesch-Kincaid grade level score is a second type of scoring system that emphasizes sentence length more than word length. Their values correspond directly to grade level. For the short stories, the grade level scores range from 3.2 to 5.8, indicating that these stories should be readable by most seventh graders. Of course, in real classrooms, reading abilities vary greatly at any particular grade level, and we do not attempt, in this book, to address the classroom-management complexity of trying to accommodate widely different abilities.

The text for individual challenges in the Learn by Example sections vary significantly in difficulty, consistent with the goal of this book to offer example challenges for a variety of experience levels. Some of these challenges are appropriate for middle school, some high school, and still others most appropriate for the teacher or expert teacher who is wanting to practice their own investigative thinking skills. Consequently, the overall average reading level for the Learn by Example investigative challenges falls significantly higher than the reading level of the stories themselves. Average Flesch-Kincaid ease of reading scores of 44.8 to 60.6 for the Learn by Example sections correspond to 9th grade through college and average Flesch-Kincaid grade level scores (corresponding directly to grade level) range from 9.3 to 11.8.

The difficulty of the Learn by Example investigative challenges themselves is not captured by a Flesch-Kincaid score or by any "ease-of-reading" type of test since the difficulty of these challenges is related to science reasoning skills, experience with investigative thinking, and prior knowledge of science. In keeping with our intention to provide encouragement and support toward deeper investigative thinking across a wide range of ability and experience levels, we offer example investigative elements that vary from challenges that most middle schoolers can engage with and learn from to others that should challenge master teachers.

If you are a teacher challenging yourself to practice investigative thinking, find the challenges right for you! If you plan to give middle or high school students some of the example investigative thinking exercises from our Learn by Example sections, we recommend that you give them only a selection of the challenges appropriate for their ability levels, or modify these challenges to work for your students.

Standing in the Shadow of Phobos

By Mary Alexandra Agner

You need a buddy to get outside the West Valles Marineris Dome, and Demetrius is having some problems in that area. Will he be able to make his first excursion outside the dome in time to watch the transit of Phobos?

Demetrius jumped onto his sister's bed, bouncing twice. He reached up to take down her wire-suspended reproduction of some old Earth building. It was the perfect place for his handmade model of Jupiter. He eyed the distance between wire and doorway: his calculations said four more centimeters to the left would be more accurate but Ann was going to be home soon from school. This would have to do.

He threaded the wire through the loop on Jupiter's north pole and tied it off. He frowned as Jupiter hung slightly sideways.

He heard the front door shut. Demetrius hopped to the floor and ran out of his sister's room, closing the door behind him.

Ann came down the hall. "Hey, junior astronomer, how was school?"

There'd been an essay but at least Teacher Anton had let him write about astronomy: tomorrow Phobos would pass directly in front of the Sun. He'd put in a paragraph about how he was celebrating his birthday by going out to watch the transit. It would be his first time outside the dome.

"Just school." Demetrius shrugged. "Will you be my Buddy so I can go out for the transit tomorrow?" None of his friends were interested and a sister for a Buddy was better than no Buddy at all.

"Maybe—" Ann reached past him for her door then stopped. She turned back to her brother. "Why are you standing here?"

Before he could answer, Ann threw the door open. "Demetrius!" She stood in the doorway, glaring back at him. Over her shoulder, Demetrius saw Jupiter hanging half-cocked.

Ann walked in, reached up, and grabbed his planet, tugging it down. Demetrius winced as the upper atmosphere tore away into filmy strands. So inaccurate. Ann shook his model—pale white and beige against her brown skin—right in his face.

"Why's this in my room?" Ann lowered her voice but it didn't matter since their parents were still at work. Mom had a surface expedition this afternoon and Dad was always doing some low-gravity plant thing.

"I built a model of the solar system. Mom said I could put the Sun in the living room. And because of where Jupiter is in its orbit now—"

"My room."

Demetrius nodded. Ann began to look resigned. Then her eyes widened. "What else did you do while you were in here?"

Demetrius glanced at her desk, giving himself away.

"My diary!" Ann screeched.

"You shouldn't use a paper one. Dad said paper is expensive—" But, Demetrius thought, much much easier to read without anyone noticing.

Ann advanced. "I'm going to get you!"

Demetrius took off running and Jupiter, propelled in a parabolic trajectory by Ann's anger, missed him, falling gently under Martian gravity to the hallway floor.

***

Dad passed Ann the dinner rolls then turned to Demetrius. "How's the studying going, son?" The smile on his dad's face was bright against his black skin. Demetrius wouldn't admit it, but he imagined Earth sunlight felt like Dad's smile. Both his parents had been born there but Demetrius and Ann had spent all their lives in the West Valles Marineris Dome.

He swallowed a bite of his dad's latest attempt at a homegrown Martian salad. "I have the test tomorrow."

"And then it's your birthday eclipse," Mom put in, her pink skin creasing when she smiled.

Demetrius said, "It's actually called a transit, Mom—"

"Keep your transits out of my room—" Ann began.

"Just a minute," Mom said, motioning Ann to wait, "I want to hear."

Demetrius said, "Both eclipses and transits have three parts: a watcher and one astronomical body passing in front of the other—from the watcher's point of view." Demetrius held up his hands, moving one in front of the other. "But with eclipses, it's more about the shadow."

"Go on," Mom said, still listening.

"Let's say I'm the watcher, on Earth, and a solar eclipse happens. The Moon—that's one astronomical body—passes in front of the Sun—that's the second body—and the Moon's shadow falls on Earth where I'm watching."

"You watch too much," Ann said with a huff. She turned to her dad. "Aren't you going to punish him for sneaking into my room?"

"Let your brother finish, Ann," Dad said quietly. Ann rolled her eyes.

"A transit happens when the astronomical body in front is smaller than the other, from the point of view of the watcher." Demetrius pulled his closer hand into a fist and smiled. "Transits are pretty special because you get to see both of the objects at the same time. And, you get to sort of feel how the entire universe is moving, when you watch a moon cross over the disk of the Sun." He wondered, were they going to laugh at him?

"So why do you need to be outside?" Ann asked. She had stopped, her fork only halfway to her mouth; Demetrius thought she'd actually been listening to him.

"It's okay if you just watch it from inside the Atrium again—" That was Overprotective Mom.

"But this is the first time I'm old enough to be outside the dome!" Demetrius had spent all year counting the days.

"Unless you don't pass the Surface Hazards test," Ann teased. That was the sister Demetrius expected.

"Ann," Dad said, his voice a warning.

Ann looked down and stabbed her fork into the green leaves.

Everyone Make Up Mom said, "Ann, honey, maybe you could go with your brother tomorrow?"

"I'm *not* going to be stuck with the junior astronaut. Not after what he did to my room today." Shaking her head, Ann left the table.

"Dad," Demetrius asked, "Can't you come with me?"

His dad sighed, then reached out and put his hand on Demetrius' shoulder. "Son, unfortunately there's a big meeting in the lab at eclipse time."

Demetrius whispered, "Transit, not eclipse," but his shoulders slumped. He turned to his mom.

"I'm sorry, Demetrius." She blushed. "But I'll already be outside. My team is on the surface tomorrow making measurements all day." She paused. "How about you help me with dishes and then I'll quiz you on sandstorm safety?"

***

The next morning, Demetrius sat in the living room balling and tossing the crumpled wrapping plastic from his birthday presents.

Today he was going to go outside the dome. He was going to stand on the surface of Mars. He was going to watch Phobos transit in front of the Sun, like he'd done every birthday since he could remember. But today, because he would be outside the dome, he was certain he would be able to feel a moment when everything was moving. He shivered in anticipation.

But to do that, he had to pass the Surface Hazards test. He brought up the tutorial on his tablet. He stared at the first question. It was about the polar geysers—many klicks south of the dome so there's no way he'd ever walk there by mistake—but he'd studied so he tapped the right answer and was rewarded with the screen blinking green.

What do you do in case of comm loss? Check with your Buddy, then head toward the dome. Tap. Blinking green screen.

When your suit indicates a high level of cosmic radiation? Check with your Buddy on their situation, run the diagnostic on the suit sensors, head toward the dome. Tap. Blinking green screen.

In case of an approaching sandstorm? Shelter in place with your Buddy. How close do you need to stay to your Buddy? How do you rate your Buddy after an expedition? Tap, tap, tap. All green.

What do you do when you want to see the transit but your family won't go with you?

Demetrius laughed out loud. You get yourself a Buddy, obviously.

He swiped off the tutorial and switched to the dome's social channels. There was already one for the transit but it didn't have a whole lot of comments. He typed rapidly, asking if anyone wanted to view it together.

The tablet began pinging almost immediately. Five people saying they wanted to go. Then another. Demetrius stared at their names, not recognizing any of them. Swiping their comments for info showed most of them to be high schoolers or college students. More pings.

The tablet screen flashed. Someone had formed a new group to view the transit together and invited him. He looked at the message and saw the group's moderator was Teacher Toni; she was much more interested in science than Teacher Anton. He stabbed the invite to accept.

He still needed a Buddy, though. Demetrius scrolled through the list of names as people accepted invites to the group viewing. He recognized some of the family names. A Larry Hernandez wanted to go. Hernandez? Juan's older brother. Juan was ok, Demetrius thought, even if he liked music more than science.

But Demetrius laughed out loud when he realized school wasn't the only place he'd heard that name recently: Larry H was in his sister's diary. Now he could go outside for the transit *and* get back at Ann for not helping him in the first place.

***

Demetrius was checking the last of the environment suit's fasteners when Ann came home from classes.

"You passed?" she asked.

Demetrius shrugged.

"Where are you going?"

Demetrius stepped back, grinning. "Out for some quality time with your boyfriend." He took off through the front door, jogging.

Ann squeaked then chased him all the way to the main air lock, stopping short when she saw the group of people already there. "What's all this?"

Demetrius picked up one of the suit helmets from the common locker and ran his fingers over it checking for defects. "None of you wanted to come with me but there are lots of other astronomers in the dome, so I asked around."

Ann looked sheepish. "I'm sorry, junior astronaut—" She held up her hands to shake away the words. "I didn't mean that—"

"I'm not a baby. I'm thirteen and I'm going out to the surface to see Phobos."

"Yeah, I know, I'm sorry I haven't been treating you more like a grown-up." Avoiding his eyes, Ann looked around again at the crowd. "That's Larry H! He's hot basalt!"

"He likes astronomy." Demetrius stuck out his tongue at his sister. "And you like him."

Ann looked at the crowd for a moment, then turned to Demetrius and smiled. So fake, he thought. "Of course I'll go with you, Demetrius."

"Too late," he said, triumphantly. "I've already got a Buddy."

***

Demetrius couldn't feel or hear the crunch of the Martian surface under his foot when he took his first step out of the airlock, but he looked down to watch the small cloud of dust rise and settle. *That's my footprint!*

Near the dome there was little to trip over, but he checked carefully before walking with Larry to the Park, a flat spot free of debris half a kilometer from the main door where the viewing group had agreed to meet. Teacher Toni was there already—she was easy to spot because of her height—with three other viewers. Artificial seats were arranged in a small arc but Demetrius stood, spinning around slowly to see every bit of the view. Larry stopped next to him. Without the windowpanes of the Atrium to interrupt, Demetrius felt as expansive as the wide red Martian sky surrounding him. He held out his hand to touch the horizon.

Demetrius checked again that Larry hadn't left—"verify the proximity of his Buddy" like the rules said—then flipped the helmet visor into polarized mode. He looked up. The suit countdown said he had nearly a minute to wait so it was just the bright disk of the sun. The creases of the suit squeezed against his skin. His neck bent back into a crick.

The sun's crisp circle of light hung motionless in the sky. As the uneven edges of Phobos began biting into it, he heard Larry and some of the others whooping over the comm channel. Demetrius smiled but didn't look away. Phobos was a smudge, then a finger-shaped blot, altering the Sun's outline. The moon was 9300 kilometers away, covering the Sun 228,000,000 kilometers away, but Demetrius held his breath anyway, caught up in the conjunction. As Phobos swept across the Sun, he felt a moment of vertigo, as though he were swinging through space along with Phobos, as Mars rotated. Bull's eye, Phobos dead center of the bright ring for barely a millisecond. Then the moon moved on, sweeping its shadow across the Martian surface. *What a birthday present I've given myself*, he thought. Demetrius inhaled deeply, then joined in the cheering over the comm channel.

***

Mary Alexandra Agner writes of dead women, telescopes, and secrets. Her poetry, stories, and nonfiction have appeared in The Cascadia Subduction Zone, Shenandoah, and Sky & Telescope, respectively. She can be found online at http://www.pantoum.org.

Art Credit: In the Shadow of Phobos artwork by Erin Colson.

"Standing in the Shadow of Phobos" was first published online in *Issues in Earth Science*, Issue 13, May 2020.

Chase Your Own Rabbits!

Question: Find something in the story that you wonder about, want to challenge, or would like to explore more fully, and ask a scientifically-investigable question about it. As you consider things you wonder about, remember that not all questions can be investigated scientifically. For example, in the story *Standing in the Shadow of Phobos*, the question "why did Demetrius want to watch the eclipse" is more literary than scientific and might be pursued through contextual clues in the story or by considering your own motivations in life. The question "how far away is Phobos from Mars" is a scientific one, but might lead to simply looking up answers rather than investigating anything. Try to find a question that can be investigated scientifically with resources that you have, or can gather, in the classroom.

The question we, Russ and Mary, address in our Learn by Example for the story *Standing in the Shadow of Phobos* is the following:

"How can you figure out what's happening out in space by watching things in the sky from the surface of a planet?"

You can choose this question as a seed-question for your own investigation or come up with your own rabbit to chase!

Feel free to write down your question here:

Vague Notion: As you prepare to plan your investigation, first get a vague notion of what you plan to do—including guidelines on whether it will involve experimentation, data-interpretation, modeling, arguing from evidence or some other practice of science.

The vague notion that we started from for our Learn by Example for this story is the following:

Go through a series of models for the solar system, including the presently accepted model, models that were thought correct in the past, and models that were always recognized as wrong even by ancient people. With this preparation, consider whether specific observations made from the surface of a planet, including observations made by Demetrius in the story, would support or undermine a particular model.

If you chose to go with our question for your own investigation, you can also choose our starting vague notion, or chase your own rabbit.

Feel free to write down your vague notion here:

Detailed Planning: Once you have your question and a vague notion in place, you can begin to fill in the nuances, complexities, and details of your plan. There is no recipe for your plan or a predetermined set of steps to follow, because your plan ***is*** the recipe or set of steps that you are creating for this particular investigation! Sketch out your investigation, anticipating as best you can what challenges you will encounter, and include as much detail as you are able (realizing that your investigative plan may not survive the first results of actually doing it.) Be sure to think about what background information you might need and how to get it, techniques you may need to research, the overall sequence of things you will need to do, and all other aspects of your investigation! <u>You will likely need several pages to document your plan.</u>

Follow a Question

Asking a question, choosing a vague notion of how to pursue that question, and then developing a detailed plan to pursue it is just the start of investigating. You actually have to do your investigation! During the course of doing your investigation, you will find out what works and what doesn't work. You may realize that you need more information, or different experimental tools, or even that your initial working model was so far off that your plan for investigation quickly falls apart. Authentic investigation involves iterative thinking during which your original experimental plan is revised, your initial interpretations updated, and your working model revisited. All of this is the whole point and practice of authentic investigation!

Keep careful notes on your experimental methods, interpretations, and models as well as your revisions as you work through your investigation.

Learn by Example (for Standing in the Shadow of Phobos)

Our Starting Question: How can you figure out what's happening out in space by watching things in the sky from the surface of a planet?

Our Starting Notion: Go through a series of models for the solar system, including the presently accepted model, models that were thought correct in the past, and models that were always recognized as wrong even by ancient people. With this preparation, consider whether specific observations made from the surface of a planet, including observations made by Demetrius in the story, would support or undermine a particular model.

Introduction

Demetrius, the young boy excited by science in the story *Standing in the Shadow of Phobos*, describes the transit of the Sun by the moon Phobos this way:

"Transits are pretty special because you get to see both of the objects at the same time. And, you get to sort of feel how the entire universe is moving."

How very true. Even before we humans had the ability to measure the distance to stars, we figured out the relative distances of objects in space and something about their motions by watching which objects pass in front of the others. We saw the Moon pass in front of the sun and the planets. We watched planets pass in front of distant stars. Galileo, with his newly-invented telescope, watched the moons of Jupiter pass in front of Jupiter, and then, a while later, watched Jupiter pass in front of the moons. With this observation, he proved that not all objects revolve around the Earth, providing the beginning of the end for the Earth-centric model for the universe.

As Demetrius points out, watching a transit can be particularly helpful in constructing a mental picture of what is going on in space. In science, we call that mental picture a *model.* Constructing models is particularly important in doing science, and is one of the important practices of science promoted by the Next Generation Science Standards (2013). For this story's Learn by Example, we are going to consider how various models for the movement of objects in our solar system can be supported or disproven by observations of the motions of bodies in space.

Remember, the following puzzles and challenges are not intended to be systematic curricular steps through which students of science work toward defined learning outcomes. Rather, they are intended as aspirational examples of investigative thinking, appropriate at a variety of levels—some middle school, some high school, some for the expert teacher—to inspire investigative thoughts and increase understanding of what investigative thinking is all about. Find one or more that is right for you!

Puzzle 1: What do Demetrius' observations tell us about the solar system?

Consider the possible solar system models below. Which of these models does Demetrius' observation of the transit disprove (select all that apply)? Explain why you think this.

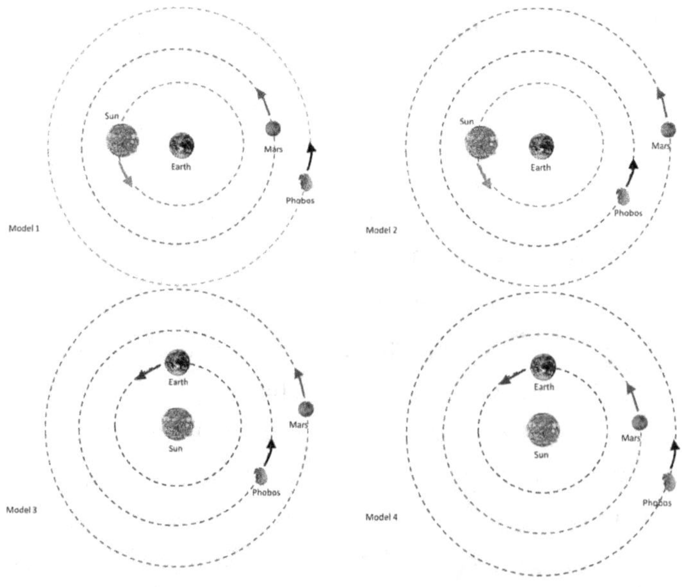

Spoiler alert: In models 1 and 4, Phobos would never come between Mars and the Sun, and Demetrius could never observe a transit of the Sun by Phobos. Demetrius' observations disprove those two models.

Now, if you have already studied the model for our solar system, you may have identified ALL of the models as being wrong. We do think they are all wrong! *But they are not all proven wrong by Demetrius' observation of the transit.*

Often in the classroom, instead of learning to identify evidence and construct models, students learn, explain, and apply the models that the scientists have given us. We just "know" which model is correct and don't always think about evidence that supports a model. Sometimes, we even make the model the evidence for itself—An example of this type of circular reasoning is when we offer an argument like "We know that model 2 is wrong because we know that the Earth orbits the Sun, not the other way around." Note that it is reasonable to apply previously established science ideas, but the idea that Earth orbits the Sun is a conclusion, not empirical evidence.

Puzzle 2: What other observations of Phobos would tell us even more?

What one additional observation of Phobos would Demetrius have almost certainly made, living on Mars, that would disprove both models 2 and 3?

If you need a hint, think about watching the sky at night. What would Demetrius see in the night sky, when the Sun is on the complete opposite side of your location on Mars? What would this observation imply?

Spoiler alert: He would have observed Phobos in the sky at night. Meaning that Phobos sometimes appears between Mars and the Sun (the transit) and Mars sometimes is located between the Sun and Phobos (when he sees Phobos in the night sky). From this, he can figure out that Phobos must orbit Mars, not only the Sun or the Earth.

Puzzle 3: Comparing the observations of Phobos to Galileo's observations of Jupiter's moons

As mentioned above, the observation that not everything in our solar system orbits the Earth was a key argument that Galileo offered against a geocentric model for our solar system. Of course, Galileo could not watch the moons of Jupiter from the surface of Jupiter, like Demetrius watches Phobos from Mars. He had to figure out that the moons orbit Jupiter by watching from Earth. Below is an illustration of the positions of the moons of Jupiter on nine sequential nights (This image is adapted from the GEMS Space Science Sequence, from the Lawrence Hall of Science, UC Berkeley).

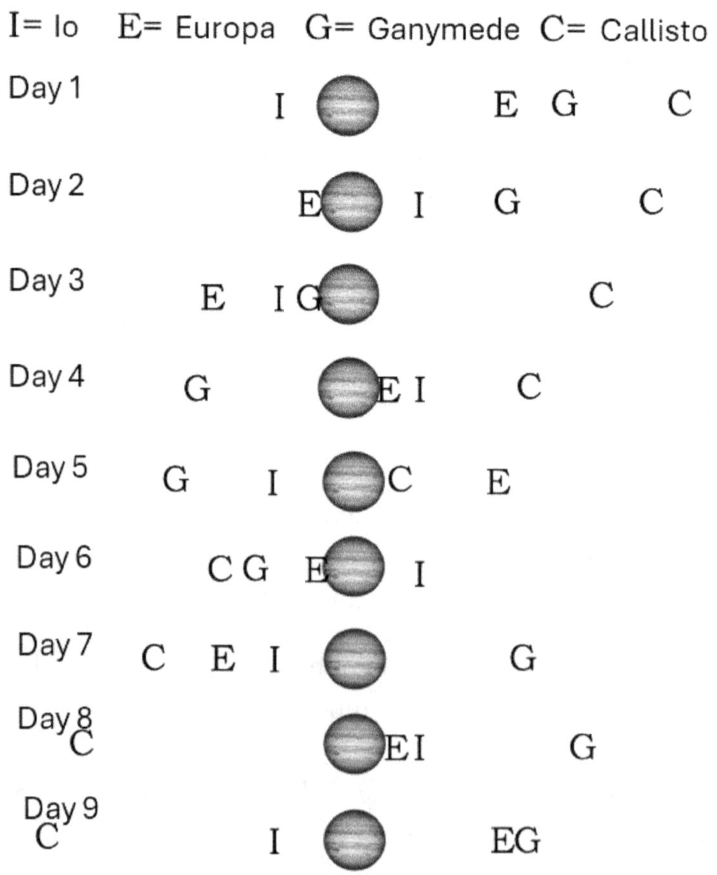

Note: the distances from Jupiter are relative, with the orbit of Callisto being about 1.17 million miles from Jupiter.

Can you figure out roughly how long it takes each moon to orbit Jupiter (Callisto, Europa, Io)? Hint: start with Callisto, the easiest one to figure out the approximate orbital period, and observe how its position changes from one day to the next, then try to get a rough estimate for the other moons' orbital periods.

Spoiler alert: You can follow the location of Callisto on one side of Jupiter, to roughly the same distance away on the other side over a period of 8 days, allowing an estimate of about 16 days for the complete trip. A more detailed analysis would be required to get to the 16.69 Earth-day value for the orbital period accepted today. A 16.69 orbital period corresponds to about 21.6 degrees each day, as shown in the illustration below.

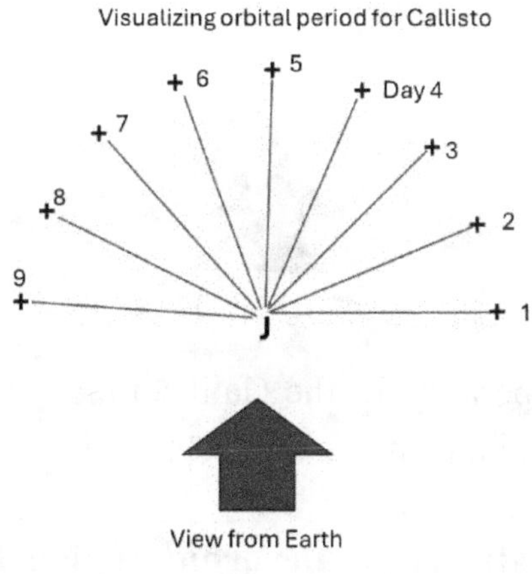

Europa is closer to Jupiter and has a shorter orbital period. On days five and nine, Europa returns to roughly the same position as it had on day one, indicating one orbit about every 4 days. Again, more careful analysis would be needed to get the 3.55 day modern value for the orbital period.

Io flips sides of Jupiter each day through the 9 day period, suggesting its orbital period is close to 2 days, not too far distant from the 1.77 day modern value (although Galileo's measurement differed from the modern value by less than 0.1%!). Unlike the case with Callisto, you can't follow the position of Io as it gradually orbits Jupiter, and so determination of an exact orbital period might require more frequent measurements than only once per day. You can see the expected (modeled) position for Io with time in the image below, using the 1.77 day period accepted from modern measurement (or about 203° each day). There appears to be some mismatch between the position modeled in this image, and the image above adapted from the GEMS Space Science Sequence, suggesting that one might be slightly wrong!

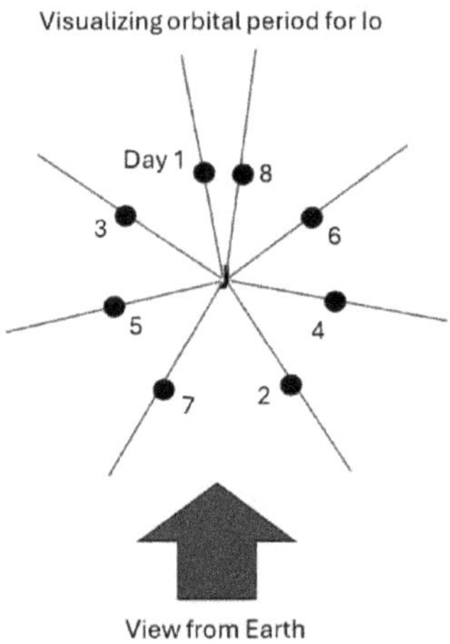

The modern orbital periods for the Galilean moons (in Earth days) are: Io= 1.77, Europa = 3.55, Ganymede = 7.15, and Callisto = 16.69.

Puzzle 4: More constraints on the orbits of planets

The following puzzle in celestial mechanics comes from *Learning to Read the Earth and Sky* (2016) by Russ and Mary Colson, published by NSTA Press.

Often people imagine that the geocentric model of Ptolemy was like that shown below. But this is not true. Even without telescopes, ancient people knew this model was not correct. Consider this model, and think about what it predicts that one could see in the night sky. What observation, which ancient people could have easily made, disproves this model?

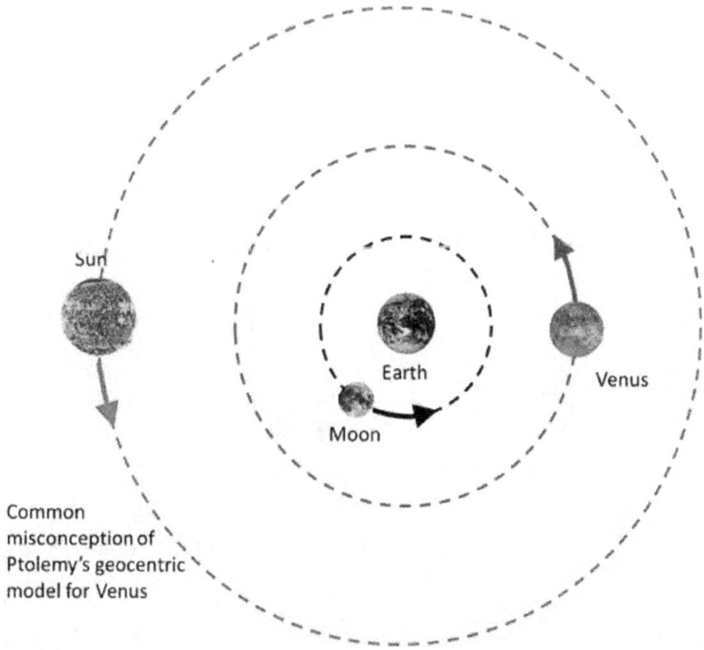

Need a hint? Think about the time of day and locations in the sky that Venus could appear.

Spoiler alert: This model implies that Venus should sometimes be seen on the complete opposite side of the Earth from the Sun. For example, we should be able to see Venus high in the sky at midnight during times when Venus has orbited to the opposite side of Earth from the Sun. In fact, Venus—the morning and evening star—is never seen at midnight from non-polar latitudes, nor is it ever seen high in the sky at night. Thus, ancient people knew this model was wrong.

A better illustration of Ptolemy's model, also from *Learning to Read the Earth and Sky* (2016), is shown below. In this model, Venus and the Sun both orbit the Earth in lock-step. Notice that this model does not allow Venus to be on the opposite side of the Earth from the Sun. The numbered positions of Venus correspond to the variations of Venus seen in the ancient sky. For example, positions 2, 3, and 4 correspond to Venus as the morning star and positions 6, 7 and 8 correspond to Venus as the evening star.

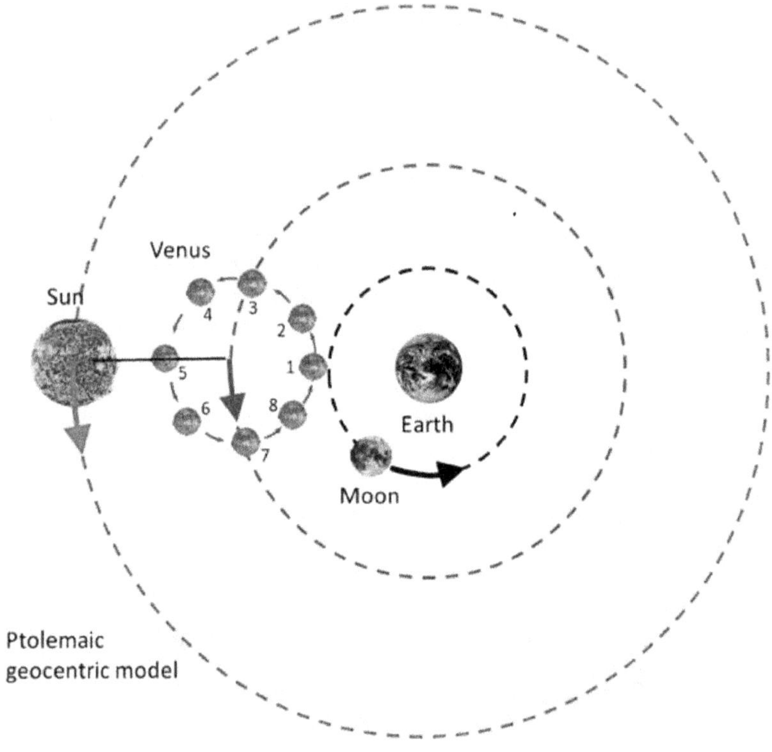

Galileo disproved this model by observing the phases of Venus through a telescope, that is, he observed how much of the lit side of Venus can be seen from Earth at different times. Only a tiny sliver of the lit side of Venus could ever be seen if Ptolemy's model were right. Yet Galileo observed a nearly fully-lit Venus, proving that Venus must orbit the Sun. Try sketching models for orbits to show that this is so. Explain your reasoning!

Further explorations of this investigative problem, with sketches of the phases of Venus expected for different orbital models, are provided in *Learning to Read the Earth and Sky* (NSTA press, 2016, pages 265-267).

Puzzle 5: What observational tests might we make to confirm the heliocentric model for our solar system if we lived on Mars?

Think about the model for the solar system shown below, showing Phobos orbiting Mars and Mars and Earth orbiting the Sun. What might the phases of Earth look like as seen from Mars? How much of the Earth would appear to be "lit" as viewed from Mars at each of the stages of Earth shown?

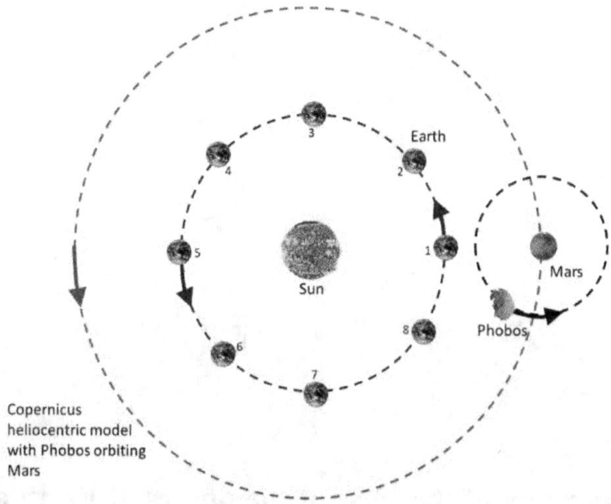

You might model the appearance of phases with reference to the illustration below.

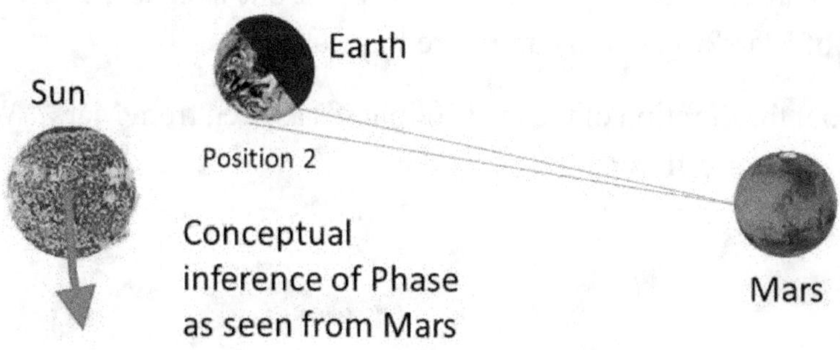

Spoiler alert: You might make a prediction of phases like that shown below.

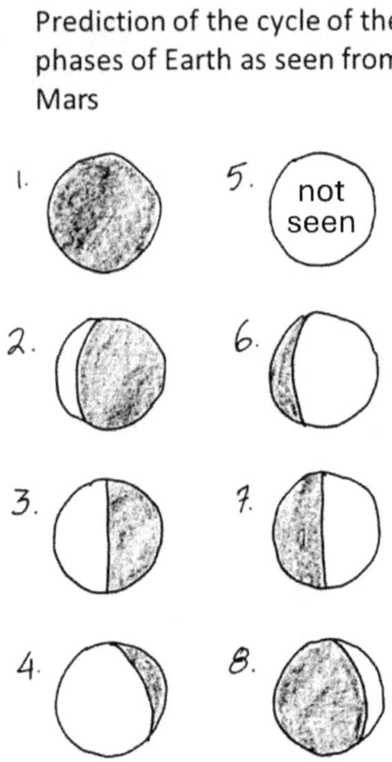

Puzzle 6: Duration of the cycle of the phases of Earth as seen from Mars

How long do you think it would take the Earth to go through its set of phases if you viewed Earth from a fixed point out in space, at about the distance of Mars? Yes, you can figure this out!

What about the duration of the cycle of phases as seen from Mars? Would it be the same or different?

Spoiler alert: Since it takes Earth one year to orbit the Sun, it will take one year to go through its set of phases as seen from a fixed location in Mars' orbit.

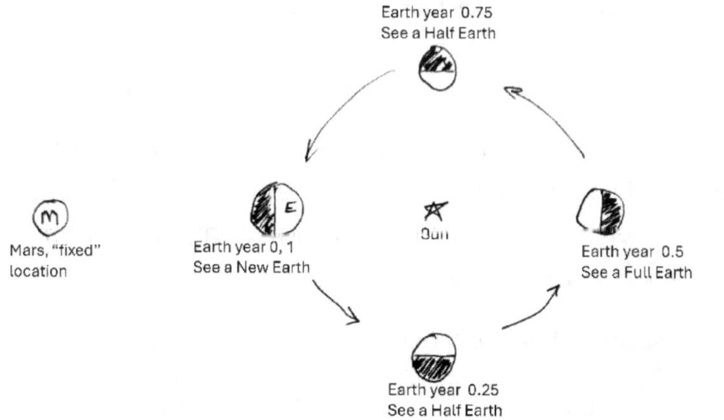

Illustration of the phases of Earth as viewed from a fixed location in Mars orbit, showing how the cycle of phases are completed in 1 Earth-year.

Spoiler alert: Since Mars is also moving as it orbits the sun, the duration of the cycle of phases of Earth as seen from Mars would be longer than one Earth year. Convince yourself by sketching out what the Earth phases would look like as Mars also orbits the Sun but not as fast as the Earth.

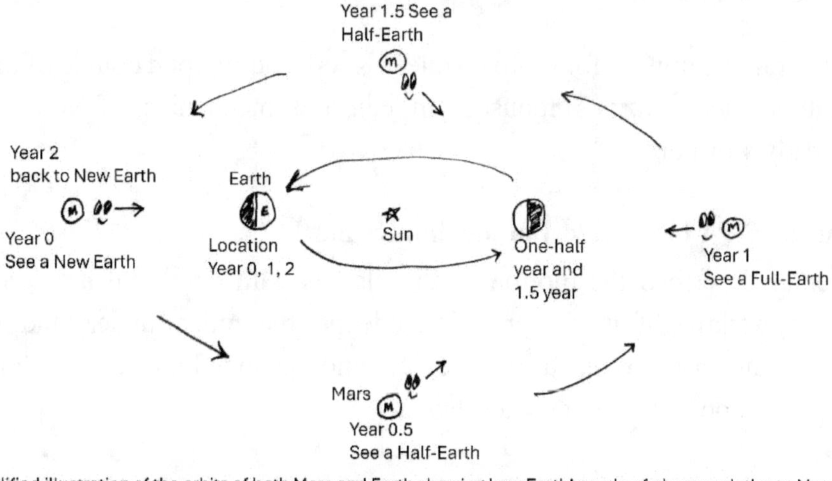

Simplified illustration of the orbits of both Mars and Earth showing how Earth's cycle of phases relative to Mars takes longer than 1 Earth-year to complete (Pretending that the Martian year is exactly twice the Earth year)

Puzzle 7: Confirming that Mars is farther from the Sun than the Earth, if you live on Earth and if you live on Mars

What would the phases of Mars look like as viewed from the Earth? Would they go through the same cycle of phases as shown above for the phases of Earth as seen from Mars, or would they be different? Sketch out what these phases would look like.

Spoiler alert: A little bit of thought reveals that since Mars' orbit is outside Earth's orbit, Mars will never appear less than half lit. There will be no "new" or "crescent" Mars! That is one way that we can confirm that Mars' orbit is outside of Earth's orbit.

What is another way that someone living on Mars could confirm this?

Think about transits!

Spoiler alert: Only planets closer than you to the Sun will be seen to transit the Sun. From Earth, it only happens with Mercury and Venus. From Mars, you could also see the Earth transit the Sun. Imagine living on Mars and watching your own home world cross the face of your sun!

Are you still hungry for more puzzles? Keep reading for a couple of more-mathematical puzzles focused on celestial mechanics, if you aren't already worn out!

Puzzle 8: A bit of celestial mechanics math

Kepler described the motion of the planets with empirical laws. Later Newton's laws of gravity provided a deeper theoretical understanding of why planetary motions are as they are and expanded the relationships to include moons and other satellites.

If one body orbits another, and if one body is much, much larger than the other, then Kepler's third law*, with Newton's modifications, can be expressed as the following:

$$P^2 \approx \frac{4\pi^2}{GM_1} a^3$$

Where P = the period of orbit, G = the gravitational constant, a = the distance between the centers of mass of the two bodies, and M_1 = the mass of the larger object. Note that this approximation ignores that M1 is actually the mass of both objects, valid in this case because the mass of Phobos is small compared to the mass of Mars.

From the story, Phobos was 9300 km from Mars. The mass of Mars is about 6.39×10^{23} kg. $G = 6.674 \times 10^{-11}\ m^3\ kg^{-1}\ s^{-2}$. Based on this, how long will it take Phobos to orbit Mars once (in Earth days)? This value will be within a few seconds of the time it takes for Phobos to go through all of its phases.

Remember, you have to get units correct. For example, did you convert kilometers to meters to match the units in G? Did you convert seconds to days to match the question request?

Spoiler alert: Period (in seconds) = sqrt(4 x 3.14 x 3.14 x 9,300,000 x 9,300,000 x 9,300,000)/(6.674E-11 x 6.39E23))

Period (in days) = Period (in seconds) /(60 x 60 x 24)

You should get about 0.316 days. Pretty fast, huh!? That little moon is really zipping around. That transit didn't last long, that's for sure!

The actual time for Phobos to go through its cycle of phases (called the synodic period) is 0.3191 days. That compares to Earth's moon at 29.5 days, nearly 100 times longer.

* Note: Kepler's three empirical laws describe the motions of the planets of our solar system. Kepler's third law relates the distance of a planet

from the Sun to the time it takes for the planet to orbit the Sun, according to the relationship $P^2 \propto a^3$.

Puzzle 9: Comparing sizes of the Moon and Sun—more math.

In the story, small Phobos does not cover the entire Sun, making it very clear that Phobos must be smaller than the Sun. That is not so obvious on Earth where the Moon can completely cover the Sun during an eclipse. Which must be larger?

Over 2000 years ago, Aristarchus measured the distances to the Sun and Moon, and measured their relative sizes compared to the size of the Earth. It was quite astonishing, at the time, to discover that the Earth was really a tiny little place in a very vast universe.

Let's reproduce one part of Aristarchus' calculation—how big is the Sun?

Aristarchus started with the observation that the Sun and the Moon appear to be about the same size in the sky (thus, the Moon just covers the Sun during an eclipse of the Sun). He first figured out the size and distance to the Moon and then the distance to the Sun (which we aren't going to do here). With this information, he calculated the size of the Sun using *similar triangles*.

Here is his concept based on the Moon and Sun appearing to be the same size in the sky:

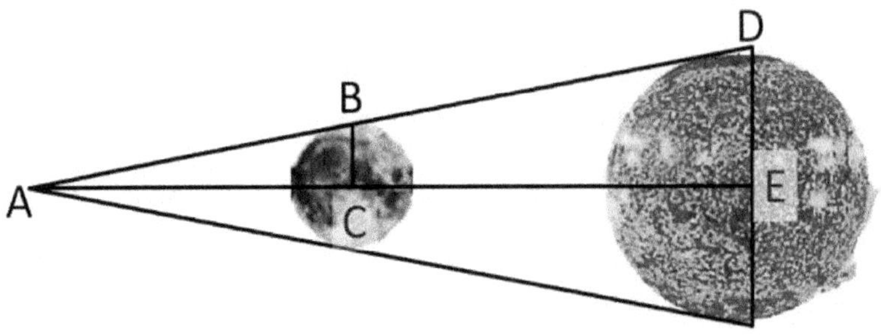

Notice that triangles A-B-C and A-D-E are similar, meaning the ratios of their sides will be equal. Knowing the size of the Moon (radius is about 0.135 Earth Diameters), the distance to the Moon (about 30.1 Earth Diameters), and the distance to the Sun (about 11679 Earth diameters). How much bigger is the Sun than the Earth (don't forget to go from radius to diameter!)

Spoiler alert: Using similar triangles, we have:

0.135/30.1 = X/11679 where X = the radius of the Sun.

X ~ 52.4, so the Sun is about 105 times bigger than the Earth! This was big news at the time of Aristarchus, when many people thought not only that the Earth was the center of the universe, but that it took up most of the space in the universe. (Aristarchus estimate was a bit smaller than this, and today it is thought that the Sun is about 109 times bigger than Earth).

Reflection on Learning

In the Introduction to *Ewww, There's some Geology in My Fiction!*, we propose that learning answers is not at all the same as doing science. Depending on what rabbits you chased during your investigation, you might have learned about important observations of the natural world or gained a deeper understanding of key theories and models that represent efforts to organize and predict those observations. However, the goal of this book is to practice investigation itself, not to focus on the information and ideas you understood better through investigation.

If traditional learning outcomes are not the goal, then what have you learned? Taking some time to consider what you gained from your investigation, reflecting on the practice of science *metacognitively*, can help you hold onto what you have gained. For example: What questions did you ask? What observations and interpretations did you make? Did you *do* modeling, rather than just learning or applying a model?

It's not easy to identify which parts of an investigation are truly investigative, both because doing an investigation ourselves is less familiar to us than learning about the results of someone else's investigative work, and because real investigation is fluid, creative, and defies easy categorization. Even so, it might be helpful to consider the following categories of investigation offered by the Next Generation Science Standard's science and engineering practices (NGSS Lead States. 2013. Next Generation Science Standards: For states, by states. Washington, DC: National Academies Press. www.nextgenscience.org/next-generation-science-standards).

- Asking questions and defining problems
- Developing and using models
- Planning and carrying out investigations
- Analyzing and interpreting data
- Using mathematics and computational thinking

- Constructing explanations and designing solutions
- Engaging in argument from evidence
- Obtaining, evaluating and communicating information

Notice that we have not numbered these practices. They are not steps in a sequential process, but rather components of an integrated and iterative process. Also recognize that this list of practices is **a** way to organize and understand the investigative practice of science, not **the** way. As you go through this reflection, be flexible in thinking about what aspects of your own investigation might be practices of science, because these practices can show up in unexpected ways. Remember, don't think so much about what you learned or understood, but rather think about what you initiated, did, practiced or experienced.

On a separate sheet of paper, jot down some things that have strengthened your ability to pursue investigative thinking. Give examples of how you engaged in particular, specific practices.

Exploring with you the experience that we, the editors, had in developing our Learn by Example section might provide some guidance as you think about your own practices of science. In the area of traditional learning outcomes, we certainly learned some things about models for the solar system and how those models developed. For us, some of this was review of things we already knew and some represented new insight into the details of the models. However, learning information or theories was not our goal. Our goal was to practice investigation. Let's consider a few examples of the investigative practices that we (the editors) experienced in developing our Learn by Example.

For the overall investigation, and for each of the individual puzzles, we had to create a question to focus our study, sometimes revising that question as our study proceeded and we gained a better understanding of where we were going (**asking questions and defining problems**).

In moving toward an answer to our starting question, we had to develop, and continually refine, an overall strategy and sequence (**planning and carrying out investigations**).

In developing the images used for the challenges, we had to read up on historical understanding of the solar system and their observational basis (**obtaining and evaluating information**).

We had to take the observational evidence, and what we read about people's understanding of those observations, and use that to create an abstract representation of what is going on in space (**developing models**).

In recognizing which wrong models are actually disproven by particular observations (such as in Puzzle 1), we had to evaluate the actual evidence, not just recall what we knew of the theories (**analyzing and interpreting data**).

In writing our spoiler alerts, we had to organize the observational evidence and connect that evidence to a reasoned conclusion (**constructing explanations**, **communicating information**, and **engaging in argument from evidence**).

In constructing Puzzle 9, we had to figure out the details of how Aristarchus determined the size and distances of the Moon and Sun, and then make choices about which aspects of his complicated calculations might be most suitable in addressing our own goals in our Learn by Example (**obtaining information**, **developing and using models**, and using **mathematics and computational thinking**).

Thus, we engaged with all the NGSS science and engineering practices, often in multiple ways. To be authentic, your investigation need not engage in all the practices. However, real investigations often include many of these practices because the practices are not separate components but are parts of an integrated whole that we call scientific investigation.

Diversion Program

By Robert Dawson

If done soon enough, the smallest diversion can save the world--or change a life.

Trigger warning: This story contains an instance of mild vulgarity and an indirect reference to race relationships.

Like a cat at a mouse hole, Major "Zig" Washington watched the UNS Beowulf's drone control board. *Ninety-nine point nine five six percent*. He ignored the straps that held him into his chair, the murmurs of his two equally-absorbed crewmates, and the drops of sweat that clung to his face and slowly grew, with no gravity to guide them away.

He raised his eyes to the viewport, completely filled by the irregular ball of rock that was the asteroid 1384756 Grendel. *Beowulf*'s orbit took ninety minutes in the asteroid's weak gravitational field; Grendel was at half phase now, but more and more of the sunlit side was becoming visible as *Beowulf* drifted past the terminator. Ninety minutes for the asteroid to go through its phases from fully lit, to dark, to lit again. *Ten kilometers across, big enough to smash all Earth back to the Stone Age if we don't change its course. Only twelve kilometers away. But we still ain't orbiting around it fast enough to earn us a speeding ticket in a school zone. Heck, I could run faster than this!* He smiled at his reflection in the heavy glass, his dark skin almost invisible against the inky black of outer space. *From Seventeenth Street to saving the world. You proud of me yet, Mama?*

The duster drones had sprayed most of the surface with crossing lines of a dull metallic shimmer, as if the rock had been wrapped carelessly in a narrow ribbon of tarnished silver. The retroreflective dust directed most of the light that hit it straight back towards the Sun, like an outfielder returning a ball to home plate. From any other direction, the sheen was subdued, like the gleam of a gun barrel. Zig's eyes dropped back to the display: ninety-nine point nine seven two percent. One hundred percent represented target coverage.

Grendel was a carbonaceous asteroid, the most common type, charcoal-grey. Astronomers had first spotted it on its last visit to the inner Solar System, seven years ago. When it was found to be an Apollo asteroid, with an orbit crossing that of Earth, they had set a computer to track it, and predict its future trajectory. The result had been so shocking that they had run the problem again, many times.

There was no doubt; the gravity of Jupiter and the gentle pressure of sunlight had, over millennia, warped the asteroid's orbit into one that would result in a catastrophic collision with

Earth in twenty-three years. They had named it Grendel then, and studied ways of diverting it. Rocket motors or atomic bombs would give only a brief push. They settled on retroreflective paint; by directing sunlight back in the direction it came from, it would almost double the light pressure. The extra push would be tiny but continuous: enough, over two decades, to move Grendel into an orbit that would miss Earth by a safe margin. Target coverage, the amount of paint needed to make this happen, was about eighty percent of the asteroid's surface. There were still irregular black polygons of bare rock between the silver ribbons.

"Almost done, Commander!" Zig said.

Commander Alina Golubova, ever graceful in free fall, appeared at his right shoulder. She reached smoothly for the grab bar on the side of his instrument panel, held it, and balanced there one-handed, swaying slowly like a plant in an aquarium. The only civilian member of the crew, Dr. Lily Zhang, a physicist specializing in orbit mechanics, hauled herself in on the other side but overshot; her shoulder bumped Zig's cheek.

"Oh! I'm sorry, Major!" She grasped the back of his chair with her other hand to stabilize herself.

"'S okay, Doc." He resumed his impassive scrutiny of the counter.

On the night side, a pale blue jet of plasma broke the utter blackness, as one of the Beowulf's duster drones adjusted its lazy hill-skimming orbit. Wherever the sun was about to rise on Grendel, that was the drones' workplace. There the electrostatic charge from the solar wind was greatest, enough to make the beads spread out and cling to the rock in a uniform coating. So the drones worked tirelessly in the predawn darkness like blind painters, spraying the beads onto the surface. As the asteroid's rotation brought the beads into the daylight, the sun's heat would fuse them permanently in place. The beads, specially engineered for this mission, cost two thousand dollars a kilogram. *Must be about the most expensive spray paint ever.*

Ninety-nine point nine nine three. C'mon mama! You can do it! Ninety-nine point nine nine eight. The glowing numbers turned from red to green. *One hundred!* He pounded his fist on the edge of the panel; his body bounced gently against his seat straps. "Commander! We have target coverage!"

"Roger, Major. Please confirm your report."

"Yes, ma'am. Retroreflective bead coverage of the asteroid Grendel is at, ah, one hundred point zero zero two percent of target value, ma'am."

"Thank you, Major." She tapped him on the shoulder, and motioned with her hand for him to turn and face the camera. "Mission Control, this is Alina Golubova, commanding the UNS Beowulf. At 0327 Standard Time, February 21, 2064…"

Zig smiled at the unnecessary formality. The Commander was talking for the history books. This one was going to be right up there with "One small step…".

"…the total coverage of retroreflective particles on the asteroid Grendel reached the target level. I repeat, we have achieved target coverage; the diversion mission has succeeded. Over and breaking for eight minutes." She switched off the video feed. In four minutes the broadcast would preempt every newsfeed on Earth, but it would be four more before Mission Control's response could reach them. Eight minutes of much-needed down time for the crew.

"Boo-yah!" Zig shouted, and punched the air.

Zhang started at the shout, then grinned. "Yes! We did it!" She pulled herself into a quick A-frame embrace with Zig, bodies almost at right angles, then did the same with Golubova.

Zig unstrapped, let himself drift free of the drone command chair, stretched as only the weightless can stretch, and wiped his face with his sleeve. *Is this what saving the world usually feels like*? he wondered. *Man, who would I even ask*? *Superman?*

"Nice work, Major!" said Golubova. She put her hand on his right shoulder and clapped him on the left, bracing so as not to put them both into a spin. "You can shut the drones off now."

"With respect, Commander, there are still some gaps. And we have plenty of dust left. Unless you want to save it for something?" He waved at the display.

"It is not necessary to spread more. That is correct, Dr. Zhang?"

"Yes, the plan already includes a twenty percent margin of error. Grendel will definitely miss Earth." She smiled wryly. "But, as Confucius puts it, 'Wherever you go, go with all your heart.'"

"What Zhang's saying, Commander, is you don't fly back to base with unexpended ammunition. No half stepping."

Zhang nodded. "Exactly. It can't hurt. It really can't."

Golubova chuckled. "Very well. Major, you may continue deploying the drones while I start preparations for departure.

But we are live again in six minutes; after I speak to Mission Control, the rest of Earth will want to hear from us. Do you perhaps have a deployment script already on the computer?" She raised one neatly-plucked eyebrow.

"It just so happens that I do, ma'am."

The Commander shook her head. "Why am I not surprised? Well, start it up."

***

Commander Golubova had made her detailed report. Mission Control had shown them brief newsclips of cheering crowds in New York, Paris, and St. Peter's Square. The interviews--live answers to prerecorded questions--were over for now, and the video link was off again. Zig looked at the viewport: Grendel was completely full, the gleaming silver spread across its face. A small, soft-edged, but dazzling disc of light surrounded Beowulf's shadow.

"Hey, peep this, peeps!" he said. "The glory's back!"

The others turned to look. "It is very bright now," said Golubova.

"The brighter, the better," said Zhang. "That means the retroreflection's effective. That's what's really going to make this work."

"Yes," said Golubova. "Though I still find it counterintuitive that all that is needed to put this asteroid onto a safe course is a layer of shiny paint. I am an old-fashioned Newtonian cosmonaut. I want to see fire behind my jets!"

Privately, Zig agreed: there was something spooky about light pressure. But he was glad the Commander had been the one to say it.

“Over twenty-three years, the acceleration will add up,” said Zhang.

“So I should take my car in for a respray and it will go faster?” Golubova asked.

“You bet, Commander,” said Zig, and grinned. “Specially if you’re still driving that Lada Elektron beater. You’d get five miles more per hour, easy, just from covering up the rust.”

“You are crazy, Zig,” Golubova said, and tapped her temple. “Choknutyy!”

“Chock nutty? That’s what my wife says, too.”

“Anyhow, it works,” said Zhang, firmly. “We make the surface retroreflective, and the acceleration almost doubles.”

“Hey, you know--maybe that’s why white folks get ahead so much easier in this world,” said Zig. “Photon pressure.” He chuckled.

Golubova’s laugh was slightly awkward. Zhang was silent.

Eventually Golubova spoke. “You have not done so badly, Zig.”

“No, I guess I haven’t. Just a moment." A line of numbers was blinking amber on the screen; he touched a few keys and the numbers vanished. “Drone C wants a course correction. Probably went too close to that high-density area near the pole. Hope the computer knows what it’s doing, because I for sure don't. Anyhow, like I was saying, for a kid who started in a street gang, I guess I've done okay."

For several seconds there was silence.

"You were in a street gang?" Golubova finally asked. "You mean, criminals? Khuligani?"

"Yeah. The Seventeenth Street Boyz. Mostly just baby gangsters. There were a few dudes who figured they were hardcore, but most of us didn't do much beyond stealing stuff and selling a bit of dope."

"Why did you do that?" asked Zhang.

"Self-defense. Mostly. You join a gang, you get a pass. Nobody trips on you, nobody beefs with your family. I don't know if I can explain to you just how it felt, coming home from school, to cross Twenty-First, see our '17' tags on the walls, and know I was in my own hood. I was safe."

"I think I understand, yes. But how'd you--get here?" Zhang asked.

"You mean, how come I ain't dead or in jail? Well, I came pretty close to both of those. See, the price of keeping our hood our own was, sometimes we had to fight for it. If another gang came and wrote their tags over ours, that was a way of dissing us, and if we let the tags stay up, we'd lose the turf. So one day we saw a bunch of Diablo tags sprayed over ours, over by the fire hall, and that night we went out to throw ours up again. And while we were writing, some Diablo assclowns came up outta the cut, caught us off guard, and started peeling caps. Two of my homies got offed, and I got winged. Then when the cops turned up, the Diablos got away, and I was the one who got caught. And I was strapped--had a gun, I mean. I hadn't even had a chance to pull it, but I was in big trouble anyhow."

"Isn't having a gun a basic human right in America?"

"Yeah, sure, Doc. Not if you're underage and the gun's hot, it ain't: I was looking at five years. But my mother spoke to the minister, and he talked to a couple of my school teachers, and they all wrote to the judge, and I got into a youth diversion program. Had to get my gang tattoo lasered, and do community service with a food bank for a year. My folks sent me to live with Aunt Jametta, in Boston. And she got me to join the Air Cadets--and here I am today, saving the world. With the capable help of you ladies, of course."

"You had to leave home?" said Golubova. "That was hard."

"Yeah. The court didn't order that, but I couldn't stay. Some of the other Seventeens were pretty mad at me--you don't just leave a gang, knamean? And, anyhow, I didn't fit in anymore. That hurt. I know we did some pretty wack stuff, but that was where I learned about being part of a team. Looking out for your homies. Stuff like that. The Air Force just had to put the polish on."

"I understand," Golubova said, quietly. Zhang was silent.

***

The Commander's voice was crisp. "All personnel should now be at stations for acceleration. Please confirm."

She's two meters away. She wants to see where we are, she can just turn her head, thought Zig. But you don't get old in this job by cutting corners. "Jet control board manned, ma'am."

"On station," said Zhang, at the astrogation board. They checked coordinates and accelerations for one last time. Finally Golubova seemed satisfied.

"Beowulf to Mission Control. This is Commander Alina Golubova. I am enabling our scheduled return flight plan. Over." She looked at Zig. "Major, help me enable takeoff." She pressed a red button on her console; he pressed its twin.

The main jets remained silent, awaiting the programmed moment; but on each console, identical red numerals appeared: minus six minutes and fifty-two seconds. The displays began to count down; a soft electronic chime repeated with each vanishing second. The chimes seemed very far apart. Zhang's knuckles were white on the arms of her seat; Golubova was impassive. For an age they sat in silence.

At minus two minutes, there was a brief hoarse bark from the attitude jets, like a sawn-off shotgun a block away, and Zig's seatback jostled his left buttock and shoulder. The stars began to crawl right-to-left across the viewport, as the Beowulf rotated into position for the big burn. *Grendel should be coming into sight real soon now.* Sure enough, a sliver of silver appeared at the right edge of the viewport, filling it second by second. Grendel was nearly full again.

"Bojemoi! What is that?" Golubova pointed at the asteroid.

The drones, executing their final script, had filled most of the polygonal voids on the gleaming face of Grendel; but they had left some areas unsprayed, leaving fuzzy irregular black lines across the silver. There, crude but recognizable, was the globe-and-olive-branches of the United Nations; and beside it, in fat bubble characters, the numeral 17.

Zig smiled, a big who-da-man grin. "Like I told you, Commander: that tag always did make me feel safer."

Golubova's response, in rich Russian idiom, was lost in the snarling roar of the main jets.

***

Robert Dawson has a PhD in mathematics from the University of Cambridge. He teaches mathematics at Saint Mary's University in Halifax, Nova Scotia. His SF stories have appeared in *Nature Futures*, *Compelling SF*, *Tesseracts XX*, and dozens of other periodicals and anthologies. His other interests include music, cycling, cooking, and hiking. He volunteers with a Scout troop. He believes that the world needs more bicycles.

Art Credit: Diversion Program artwork by Erin Colson.

"Diversion Program" was first published online in *Issues in Earth Science*, Issue 1, July 2014.

Chase Your Own Rabbits!

Question: Find something in the story that you wonder about, or want to challenge, or would like to explore more fully, and ask a scientifically-investigable question about it. As you consider things you wonder about, remember that not all questions can be investigated scientifically. For example, in the story *Diversion Program*, the questions "is asteroid 1384756 Grendel a real asteroid" and "what is a glory" are simply factual-information questions addressed with a Google or AI search and do not really engage with investigative thinking. You might ask a more investigable question (like "what processes of light cause a glory") and then try to create a model or do an experiment to figure it out. A question like "Is ten kilometers across really big enough to smash all Earth back to the Stone Age like the story claims" can involve investigation, like calculating the energy released at impact or discussing whether or not that would put people back to the stone age, although it might prove difficult to base the discussion on science and not on preconceived notions. Try to find a question that can be investigated scientifically with resources that you have, or can gather, in the classroom.

The question we, Russ and Mary, address in our Learn by Example for the story *Diversion Program* is the following:

"How do we model the motion of objects in space, like the movements of the Beowulf *and Grendel implicit in the story or the movements of our own Earth and Moon?"*

You can choose this question as a seed-question for your own investigation or come up with your own rabbit to chase! There are actually a lot of question possibilities in this story, including questions about solar wind, electrostatic charge, retroreflectivity, light pressure, glories, and carbonaceous asteroids, among others.

Feel free to write down your question here:

Vague Notion: As you prepare to plan your investigation, first get a vague notion of what you plan to do—including guidelines on whether it will involve experimentation, data-interpretation, modeling, arguing from evidence or some other practice of science.

The vague notion that we started from for our Learn by Example for this story is the following:

Consider our model for the phases of Earth's Moon, and then apply that modeling to understanding what happens in Diversion Program with Grendel and the Beowulf. *Consider the mathematics of a variety of motion events in the story to better understand the story and its context.*

If you chose to go with our question for your own investigation, you can also choose our starting vague notion, or chase your own rabbit.

Feel free to write down your vague notion here:

Detailed Planning: Once you have your question and a vague notion in place, you can begin to fill in the nuances, complexities, and details of your plan. There is no recipe for your plan or a predetermined set of steps to follow, because your plan ***is*** the recipe or set of steps that you are creating for this particular investigation! Sketch out your investigation, anticipating as best you can what challenges you will encounter, and include as much detail as you are able (realizing that your investigative plan may not survive the first results of actually doing it.) Be sure to think about what background information you might need and how to get it, techniques you may need to research, the overall sequence of things you

will need to do, and all other aspects of your investigation! You will likely need several pages to document your plan.

Follow a Question

Asking a question, choosing a vague notion of how to pursue that question, and then developing a detailed plan to pursue it is just the start of investigating. You actually have to do your investigation! During the course of doing your investigation, you will find out what works and what doesn't work. You may realize that you need more information, or different experimental tools, or even that your initial working model was so far off that your plan for investigation quickly falls apart. Authentic investigation involves iterative thinking during which your original experimental plan is revised, your initial interpretations updated, and your working model revisited. All of this is the whole point and practice of authentic investigation!

Keep careful notes on your experimental methods, interpretations, and models as well as your revisions as you work through your investigation.

Learn by Example (for Diversion Project)

Our Starting Question: "How do we model the motion of objects in space, like the movements of the Beowulf *and Grendel implicit in the story or the movements of our own Earth and Moon?"*

Our Starting Notion: Consider our model for the phases of Earth's Moon, and then apply that modeling to understanding what happens in Diversion Program with Grendel and the Beowulf. Consider the mathematics of a variety of motion events in the story to better *understand the story and its context.*

Introduction

Our starting question for *Diversion Project* addresses ideas of motion of objects in space and how that understanding can inform our understanding of the story. From *Diversion Project* we read that

"Grendel was at half phase now, but more and more of the sunlit side was becoming visible as Beowulf *drifted past the terminator."*

What does this mean? Why would Grendel be at half-phase as the *Beowulf* drifted over the terminator? Why would the sunlit side become more visible with time?

To answer these questions, and to understand the story, we have to visualize what is being described. Addressing these questions and understanding the story requires significant mental modeling of phases and motions in space!

Regardless of our experience or knowledge in science, we always start a new investigation somewhere in the middle of it—not at the point of knowing nothing and not at the point of complete understanding. For our consideration of what is going on at Grendel, let's start with the presumption that we understand the basic idea of what causes day and night (a rotating earth bathed by sunlight) and phases of the Moon (an orbiting Moon bathed by sunlight). Just to get ourselves thinking about terminators, rotations, and phases on Grendel, let's first practice some exercises with Earth's day-night cycle and the Moon's phases, with an emphasis on figuring out which direction the Earth must be rotating, and which direction the Moon must be orbiting.

Remember, the following puzzles and challenges are not intended to be systematic curricular steps through which students of science work toward defined learning outcomes. Rather, they are intended as aspirational examples of investigative thinking, appropriate at a variety of levels—some middle school, some high school, some for the expert teacher—to inspire investigative thoughts and increase understanding of what investigative thinking is all about. Find one or more that is right for you!

Challenge 1: Figuring out Earth's rotational direction through modeling and observation

Think about our model for the daylight and nighttime sides of the Earth as bathed by the Sun. To help visualize, you might set up something like what is shown below, using a small globe for the Earth and a flashlight for the Sun.

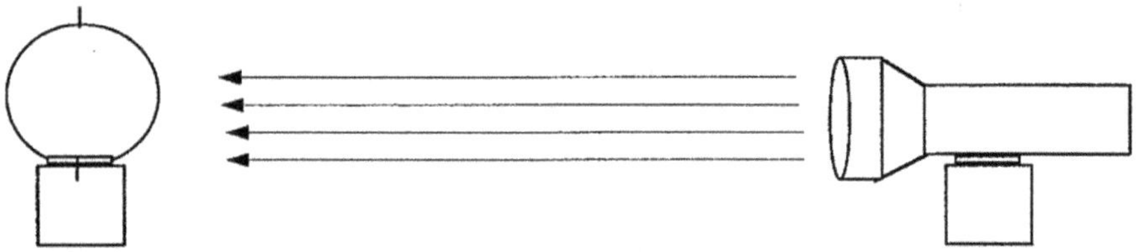

The rotational axis is indicated by the short lines at the top and bottom of the Earth globe (for our simple modeling, we are going to ignore the fact that the Earth is tilted 23.5° relative to the line of incoming sunlight).

PUZZLE 1: From this perspective, which of the following do we expect to be true (select all correct)?

- ☐ The terminator (the boundary between light and dark) will run vertically from top to bottom
- ☐ The terminator (the boundary between light and dark) will run horizontally from left to right
- ☐ Night time will be at the top, day at the bottom
- ☐ Night time will be at the bottom, day at the top
- ☐ Night time will be to the left, day to the right
- ☐ Night time will be to the right, day to the left

Spoiler alert: We can see that the right side of the Earth toward the flashlight will be lit, the left side away will be dark, and the boundary between them will run vertically up the middle.

PUZZLE 2: Now let's change our perspective and look down on our Earth model as though from above the North Pole, as in the illustration below. Consider the location of the terminator and lit and dark sides again (and again ignore the tilt of Earth's rotational axis relative to incoming sunlight).

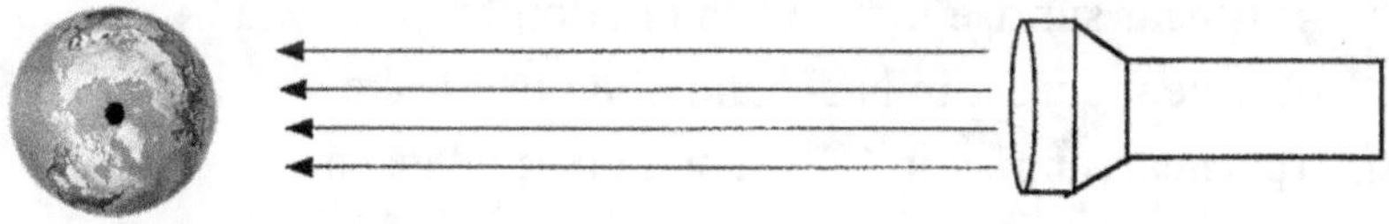

Zooming in a bit (below), find where New York City would be on the east coast of North America.

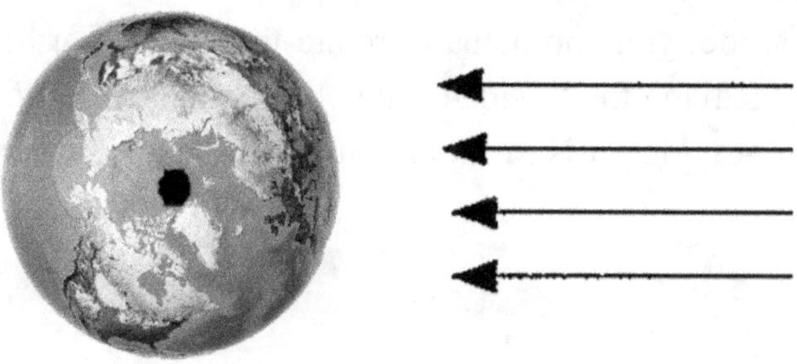

Suppose that you are standing in New York City, looking for the Sun. At the time shown in the model image, where would you look to see it?

a. West, over the North American continent
b. North, toward the North Pole
c. East, across the Atlantic Ocean
d. South, toward the equator

Spoiler Alert: Based on the direction the sunlight is coming from, as seen in the model above, the sun will be coming over the Atlantic Ocean, which we say is to the east of New York City. Notice that relative to a perspective from out in space, east appears to be different direction in different locations around the Earth!

PUZZLE 3: Suppose that the Earth is rotating clockwise, as seen from this perspective. Based on the model, which of the following can we infer?

a. It is near sunset in NYC and midnight in Alaska
b. It is near sunrise in NYC and midnight in Alaska
c. It is near sunset in NYC and noon in Alaska
d. It is near sunrise in NYC and noon in Alaska

Spoiler Alert: Setting the Earth to rotating in our imagination and using the model above (or, if you have set up a physical model with a flashlight and small globe, you can actually rotate the model Earth), we see that NYC is just leaving the lit side of the planet and going into night, which we call sunset. Alaska is in the middle of the night side, which will be midnight.

PUZZLE 3: Suppose that the Earth is rotating <u>counterclockwise</u>, as seen from this perspective.

Based on the model, which of the following can we infer?

a. It is near sunset in NYC and midnight in Alaska
b. It is near sunrise in NYC and midnight in Alaska
c. It is near sunset in NYC and noon in Alaska
d. It is near sunrise in NYC and noon in Alaska

Spoiler Alert: Setting the Earth to rotating in our imagination and using the model above (or, if you have set up a physical model with a flashlight and small globe, you can actually rotate the model Earth), we see that NYC is just entering the lit side of the planet and going into day, which we call sunrise. Alaska is in the middle of the night side, which will be midnight.

PUZZLE 4: Which of the following statements is the most-correct assessment of what we can conclude?

a. We know that Earth is spinning counterclockwise, as viewed from this perspective, because the model shows this to be the better interpretation
b. We know that Earth is spinning counterclockwise, as viewed from this perspective because we observe that the sun is always seen in the east at sunrise in NYC
c. We can't actually know which direction the Earth is spinning from this analysis because this is just a model, not an actual experiment or observation
d. We can't actually know which direction the Earth is spinning from this analysis because it will be completely the opposite if we were to view the Earth from below the South Pole.

Spoiler Alert: A model alone cannot prove any particular conclusion, which, to be science, must be based on observation. Observations that prove that the Earth is rotating counterclockwise from this perspective include the sun always rising in the east, and daylight starting earlier in farther-eastward locations. Because we can all make these observations, we can conclude that the Earth is rotating counterclockwise, as viewed from the north, based on the observations (not on the model "showing this is a better interpretation"). Of course, as viewed from the South Pole, the rotation would appear to be clockwise, but the direction of Earth's rotation from the perspective shown in the diagrams would still be counterclockwise.

PUZZLE 5: Consider the composite visualization below from NASA Earth Observatory, showing Europe in shadow, with England, Scotland, Ireland, Spain, and part of Africa in sunlight.

From our modeling and observations, which of the following can we conclude?

a. England and Ireland will soon be experiencing sunrise
b. England and Ireland will soon be experiencing sunset
c. England and Ireland have just experienced sunrise
d. England and Ireland have just experienced sunset

Spoiler Alert: Given the rotation direction of the Earth, we expect the terminator to progress from east to west, right to left in this image. In other words, Earth will rotate into the darkness. Thus, this terminator represents sunset, which England and Ireland will soon experience. If the Earth rotated in the other direction, then we would conclude that England and Ireland had recently experienced sunrise!

Challenge 2: Figuring out the Moon's orbital direction through modeling and observation

Let's add the Moon into our model. You might get a small globe on a stick, perhaps painted half black and half white, to place at different locations around your Earth globe to model the Moon's orbit, such as shown below from a perspective above Earth's North Pole.

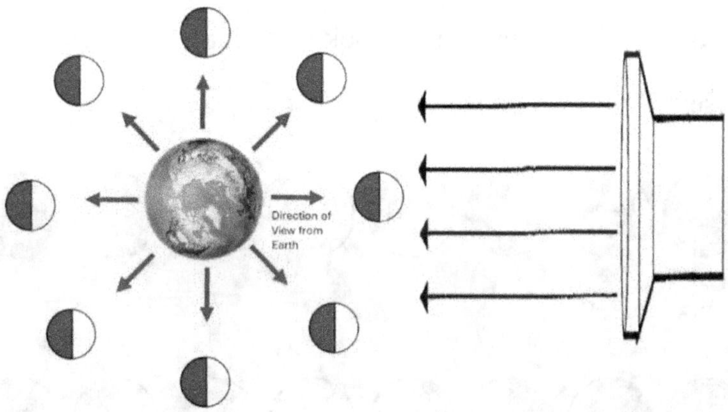

By conceptualizing a view from the Earth, or peering over the model globe of the Earth if you have built the physical model, you can construct what the lit part of the Moon looks like at different orbital positions (as viewed from the Northern Hemisphere).

PUZZLE 6: Suppose that the Moon orbits the Earth in a <u>clockwise</u> fashion. Visualize (or peer over the Earth at the Moon lit by your distant flashlight in a darkened room) to create the sequence of Moon phases that you observe for each of the positions shown in the image above (start with New Moon, when the Moon is between the Sun and Earth)

Now do the same exercise, but presuming that the Earth orbits in a counterclockwise fashion.

Can you tell the difference?

Spoiler Alert: Your images might look something like this:

Clockwise:

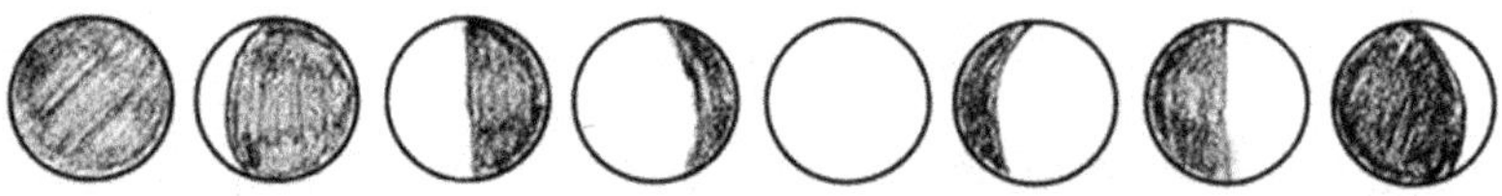

Counterclockwise:

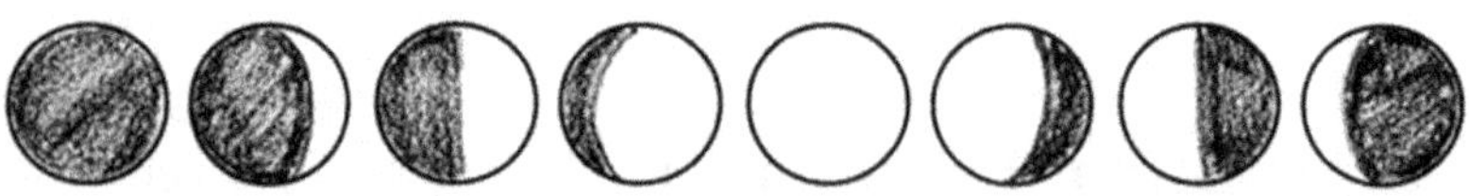

PUZZLE 7: So, does this modeling tell us which way that the Moon orbits? No! We still need real-world observational data to tell that!

Below are the observed phases of the Moon, as seen from the Northern Hemisphere, over 28 days in 2005.

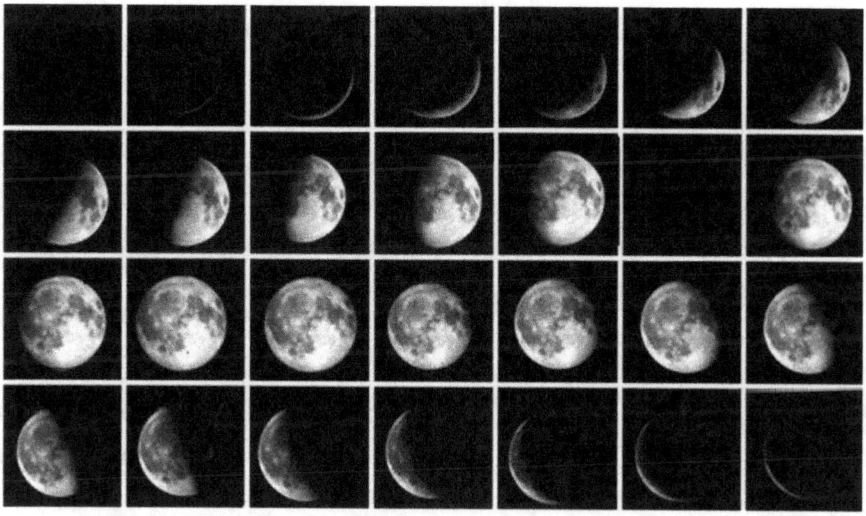

From these observational data, plus our modeling, which of the following can we conclude?

a. The Moon orbits counterclockwise as viewed from above the North Pole
b. The Moon orbits clockwise as viewed from above the North Pole
c. Since the real Moon phases don't exactly match any of our models, we can conclude that our model itself is wrong.
d. Since the terminator of the Moon is not oriented up and down, we can conclude that the Moon is tilted in space much more than the Earth

Spoiler Alert: The sequence of phases matches the case where the Moon orbits counterclockwise as viewed from above the North Pole and the crescent with the right side lit appears first after New Moon, and then the left side is lit just prior to return to New Moon. It is certainly true that the terminator is not straight up and down in the phase calendar and that observation does not match our modeling. But we already knew that we had made simplifying assumptions in our model, such as ignoring Earth's tilt and we also ignored the latitude on Earth we view the Moon from, both of which affect the apparent orientation of Moon's terminator. However, neither of these differences meaningfully undermine the more basic relationship of the sequence of phases. Finally, the tilt of the Moon's axis cannot affect the parts of the Moon that we see lit as viewed from the Earth.

Challenge 3: Grendel and the *Beowulf*

We can develop similar models for the motion of Grendel and the *Beowulf.* We might use a small potato to represent Grendel and draw on the potato with a black marker to shade the dark side of the asteroid that faces away from the Sun. The terminator is then represented by the boundary between normal and blackened parts of the potato-turned-asteroid (although shading a dark side does not allow us to see how the terminator moves across the surface of the asteroid as the asteroid rotates). We can sketch out on paper what Grendel looks like as seen from the *Beowulf* at different positions in the *Beowulf's* orbit around Grendel. For additional spatial visualization, we can sketch what both Grendel and the *Beowulf* look like at those same positions, if viewed from a distant fixed location.

PUZZLE 8: Let's consider the quote that we started from:

Grendel was at half phase now, but more and more of the sunlit side was becoming visible as Beowulf *drifted past the terminator."*

Can you figure out how the Sun, Grendel, and the *Beowulf* must be oriented in space at this moment in the story? Indicate EACH of the following that fits the description in the story ("B" indicates the position of the *Beowulf*, and "G" the position of the Grendel.

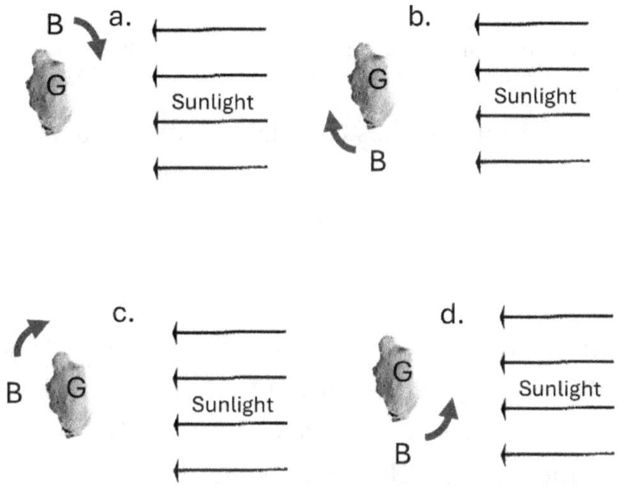

Spoiler Alert: The *Beowulf* is drifting over the terminator and that means *Beowulf* is between the shadowed side of Grendel and the sunlit side, eliminating the option in the lower left. Also, the ship is travelling into the sunlit side, so the upper right doesn't work either. The other two are equivalent, and could simply be the same scene viewed from different sides of the asteroid, like viewing phases of the Moon from the southern or northern hemispheres.

PUZZLE 9: Let's look at a bigger chunk of the description above.

He raised his eyes to the viewport, completely filled by the irregular ball of rock that was the asteroid 1384756 Grendel. Beowulf*'s orbit took ninety minutes in the asteroid's weak gravitational field; Grendel was at half phase now, but more and more of the sunlit side was becoming visible as* Beowulf *drifted past the terminator. Ninety minutes for the asteroid to go through its phases from fully lit, to dark, to lit again.*

Why does it take 90 minutes for Grendel to go through its cycle of phases?

a. That is how long it takes Grendel to rotate once on its axis of rotation
b. That is how long it takes Grendel to orbit its home planet or moon
c. That is how long it takes Grendel to orbit the Sun
d. That is how long it takes the *Beowulf* to orbit Grendel.

Spoiler alert: Although the rotation and tumbling of Grendel will affect which part of Grendel is illuminated at any point in time, the terminator, from the point of view of the *Beowulf*, is only determined by the orientation of the asteroid relative to the Sun and the *Beowulf*'s location. Thus, the *Beowulf*'s orbit rate determines how quickly one viewing from the *Beowulf* sees the cycle of phases.

PUZZLE 10: Case 1. If someone was standing on Grendel, say at the terminator, where would the Sun and *Beowulf* appear in the sky if Grendel was at full phase as seen from the *Beowulf*?

a. Case 1: Sun on the horizon and *Beowulf* overhead
b. Case 1: Sun overhead and *Beowulf* on the horizon
c. Case 1: Sun near the horizon, and *Beowulf* even closer to the horizon
d. Case 1: Sun near the horizon and *Beowulf* a bit farther from the horizon

Case 2. Where would Sun and *Beowulf* appear in the sky if you were at "midday" on Grendel and the phase was first or third quarter as seen from the *Beowulf*?

a. Case 2: Sun overhead and *Beowulf* on the horizon
b. Case 2: Sun on horizon and *Beowulf* overhead
c. Case 2: Sun overhead and *Beowulf* overhead
d. Case 2: Sun on horizon and *Beowulf* on the horizon

Spoiler Alert: Case 1. Standing on the terminator means that the sun is either rising or setting from the POV of you on Grendel, thus the sun will be on the horizon for Case 1. Because Grendel is at full phase from the point of view of the *Beowulf*, we know that the *Beowulf* is over the "noon" point of Grendel, and so in the same direction as the Sun. Because *Beowulf* is closer to Grendel than the Sun, it will appear even closer to the horizon than the Sun as viewed from the surface of Grendel, as shown in the picture below.

Case 2. Midday on Grindel means the sun will be overhead for Case 2. If Grendel appears to be first or third phase from the *Beowulf*, then the *Beowulf* must be over the terminator, and so would appear on the horizon to someone directly under the Sun, as shown in the picture.

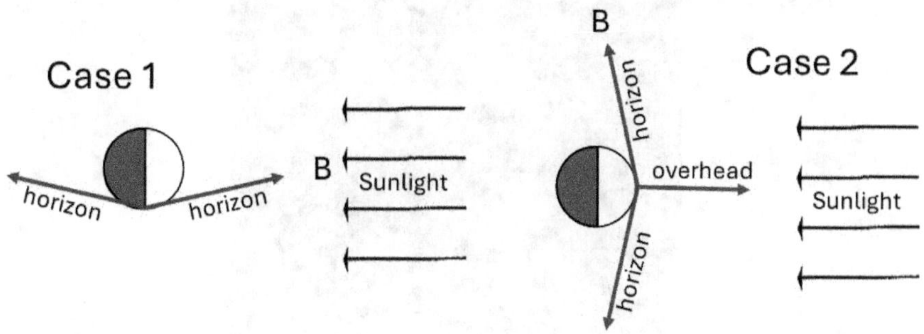

PUZZLE 11: If you were standing on the surface of Grendel, the terminator would move relative to your location because of which factors?

a. The rotation of Grendel
b. The orbit of the *Beowulf*
c. Both the rotation of Grendel and the orbit of the *Beowulf*
d. Neither the rotation of Grendel or the orbit of the *Beowulf*

Spoiler Alert: From the point of view of someone on Grendel, the location of the terminator will move as Grendel rotates in the sun, and not be related to the orbit of the *Beowulf*, just like the terminator on Earth (sunrise and sunset) depends on Earth's rotation and not the position of our Moon, or space stations, or any other satellites orbiting the Earth.

PUZZLE 12: Artists often use some artistic license in interpreting their subject. The illustration for *Diversion Program* shown below looks like it could represent the moment when the rest of the crew spots Zig's artwork on Grendel's surface at the end of the story (notice the dark number 17 in the picture). However, the picture doesn't quite match this moment. Using your understanding of phases, explain why not (checking back with the original text in the story). Image credit: Erin Colson.

Spoiler Alert: In the story, Grendel is at full phase relative to the perspective of the *Beowulf*, and therefore this dramatic view of the bright sunlight just beyond the terminator would not be possible.

Challenge 4: Some mathematical thinking

PUZZLE 13: The title of the story refers to how a small diversion "in either a life or an asteroid" can produce a big effect. However, light pressure surely can't move an asteroid by very much, right? Can you explain how it might be enough for the asteroid to miss Earth entirely?

Spoiler Alert: Think about two straight paths that go at slightly different angles. To start with, the paths stay very close together. But as they go farther and farther from the starting point, they diverge significantly. Grendel will travel hundreds of millions of miles before it encounters Earth, so a very small diversion causes a big difference in where it is at the moment it crosses Earth's orbit in 23 years. Consider the simplified conceptual illustration of paths shown below, remembering that both the asteroid and the Earth are moving (Earth orbits the Sun many times as the asteroid approaches, and, although it is not specified in the story, Grendel also might orbit the Sun several times during those 23 years).

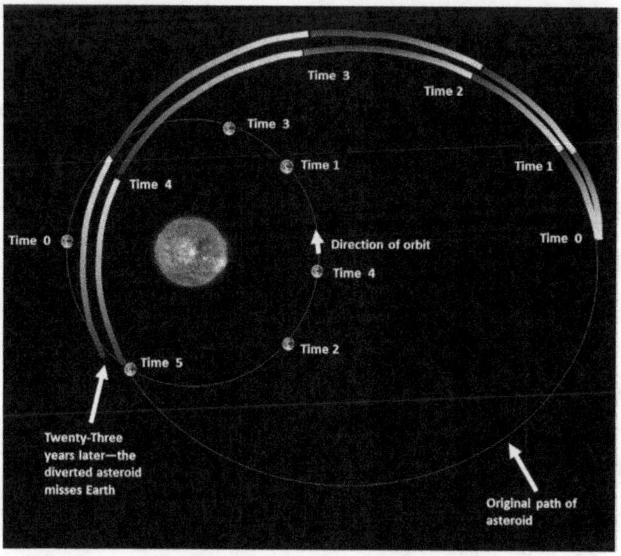

PUZZLE 14: At one point in the story, Zig comments that they have reached 100.002% of target coverage of the asteroid.

Is more than 100% possible? What if the story reported 100.002% *coverage* rather than 100.002% of *target* coverage?

Are there situations where greater than 100% might not be possible? For example, if someone reported that a rock was made of 100.002% SiO_2, is that possible?

Spoiler Alert: They reached more coverage than they had originally planned, and so 100.002% of targeted coverage makes perfect sense. You can't, in general have a chemical component that makes up more than 100% of something, however, sometimes concentrations of greater than 100% or less than 0% reflect the intrinsic analytical uncertainties of measurements. If the uncertainty of a measurement is plus or minus 1% for example, a value of -1% or 101% becomes possible analytically, although it would not mean that the material really was less than nothing or more than everything.

PUZZLE 15: Zig mentions in the story that his speed in orbiting Grendel is not "fast enough to earn us a speeding ticket in a school zone."

Let's calculate his speed given how long it takes to orbit the asteroid, the size of the asteroid, and the *Beowulf's* distance above the surface. To do this calculation, we need to know the relationship between diameter and circumference of a circle and be aware of the idea that velocity is a distance traveled divided by the time to travel that distance.

Here is the key part of the text:

Ninety minutes for the asteroid to go through its phases from fully lit, to dark, to lit again. Ten kilometers across, big enough to smash all Earth back to the Stone Age if we don't change its course. Only twelve kilometers away. But we still ain't orbiting around it fast enough to earn us a speeding ticket in a school zone.

Spoiler Alert: The radius of the *Beowulf*'s orbit can be estimated by half the diameter of the asteroid, plus the distance above its surface = 5km + 12 km = 17km. Circumference, the distance traveled by the *Beowulf* = 2 x radius x π, = 2 x 17km x 3.1416 = 106.8km, which they travel in 90 minutes, or 1.5 hours, so they are travelling at about 71.2km/hour or 44.2 miles/hour. Maybe you'd get a ticket in a school zone at that speed?

PUZZLE 16: In the story, it takes four minutes for a message to go from the *Beowulf* to Earth and another four minutes for an answer to return from Earth. Based on the speed of light in vacuum, 186,000 miles per second, how far is the *Beowulf* from Earth?

Spoiler Alert: 4 minutes x 60 seconds/minute x 186,000 miles/sec = 44,640,000 miles. This means that at the time of the expedition, Grendel was inside the orbit of Mars!

Reflection on Learning

The goal of this book is to practice investigation itself, not to focus on the information and ideas you understood better through investigation. If traditional learning outcomes are not the goal, then what have you learned? Taking some time to consider what you gained from your investigation, reflecting on the practice of science *metacognitively*, can help you hold onto what you have gained. For example: What questions did you ask? What observations and interpretations did you make? Did you *do* modeling, rather than just learning or applying a model?

Real investigation is fluid, creative, and defies easy categorization. Even so, it might be helpful to consider the following categories of investigation offered by the Next Generation Science Standard's science and engineering practices (NGSS Lead States. 2013. Next Generation Science Standards: National Academies Press).

- Asking questions and defining problems
- Developing and using models
- Planning and carrying out investigations
- Analyzing and interpreting data
- Using mathematics and computational thinking
- Constructing explanations and designing solutions
- Engaging in argument from evidence
- Obtaining, evaluating and communicating information

Notice that we have not numbered these practices. They are not steps in a sequential process, but rather components of an integrated and iterative process. Also recognize that this list of practices is **a** way to organize and understand the investigative practice of science, not **the** way. As you go through this reflection, be flexible in thinking about what aspects of your own investigation might be practices of science, because these practices can show up in unexpected ways. Remember, don't think so much about

what you learned or understood, but rather think about what you initiated, did, practiced or experienced.

On a separate sheet of paper, jot down some things that have strengthened your ability to pursue investigative thinking. Give examples of how you engaged in particular, specific practices.

Exploring with you the experience that we, the editors, had in developing our Learn by Example section might provide some guidance as you think about your own practices of science. In the area of traditional learning outcomes, we certainly reviewed the causes of day and night on Earth and the phases of the Moon. We had to dig deeper into understanding how Earth's tilt and the tilt of the Moon's orbit around the Earth interact with each other to cause subtle variations in the appearance of the Moon in the sky (although much of this did not, in the end, make it into our Learn by Example investigation). However, learning information or theories was not our goal. Our goal was to practice investigation. Let's consider a few examples of the investigative practices that we (the editors) experienced in developing our Learn by Example.

A key focus of the challenges on Earth's rotation and Moon phases (Challenges 1 and 2) is that models are not disembodied truths given to us by scientists, but rather they are based on observations. This focus guided us toward the sequence and conceptual steps of our investigation (our **planning and carrying out an investigation**). A key **question** we had to ask ourselves was what, exactly, are those observations and how do they lead us to the correct understanding? Then, we had to **analyze** the actual observed phases of the Moon in Puzzle 7 in order to relate them, through **modeling**, to an **interpretation** of which way the Moon orbits.

Whereas we **developed models** in the exercises of Challenges 1 and 2, we **applied those models** to exercises in Challenge 3. We connected the different models, and established the validity of the models, through **constructing explanations** and **arguing from evidence**, including in the Spoiler Alerts.

We engaged in some **mathematical thinking** in Challenge 4. Mathematical thinking is not accomplished by punching numbers into a calculator. The mathematical thinking practice involves figuring out how to *set up* the calculation. This involves substantial **modeling** of the scenario to figure out what values are needed and what mathematical manipulations can get us from what we know to what we are trying to **compute**. We had to get additional information to set up the problem and do these computations which we did by **obtaining and evaluating information**, including the speed of light, formula for circumference of a circle, and others.

Thus, we again engaged with all the NGSS science and engineering practices, often in multiple ways. To be authentic, your investigation need not engage in all the practices. However, real investigations often include many of these practices because the practices are not separate components but are parts of an integrated whole that we call scientific investigation.

Jigsaw

By Douglas Smith

Cassie Morant loves puzzles, but can she put together the planetary geology clues fast enough to save the landing team from execution?

Still in shock, Cassie Morant slumped in the cockpit of the empty hopper, staring at the two viewplates before her.

In one, the planet Griphus, a blue, green and brown marble wrapped in belts of cloud, grew smaller. Except for the shape of its land masses, it could have been Earth.

But it wasn't. Griphus was an alien world, light-years from Sol System.

A world where nineteen of her shipmates were going to die.

And one of them was Davey.

On the other viewplate, the segmented, tubular hull of the orbiting Earth wormship, the *Johannes Kepler*, grew larger. Cassie tapped a command, and the ship's vector appeared, confirming her fears.

The ship's orbit was still decaying. She opened a comm-link.

"Hopper two to the *Kepler*," she said. "Requesting docking clearance."

Silence. Then a male voice crackled over the speaker, echoing cold and metallic in the empty shuttle. "Acknowledged, Hopper two. You are clear to dock, segment beta four, port nine."

Cassie didn't recognize the voice, but that wasn't surprising. The *Kepler* held the population of a small city, and Cassie was something of a loner. But she had no trouble identifying the gruff rumble she heard next.

"Pilot of hopper, identify yourself. This is Captain Theodor."

Cassie took a breath. "Sir, this is Dr. Cassandra Morant, team geologist."

Pause. "Where's team leader Stockard?" Theodor asked.

Davey. "Sir, the rest of the surface team was captured by the indigenous tribe inhabiting the extraction site. The team is..." Cassie stopped, her throat constricting.

"Morant?"

She swallowed. "They're to be executed at sunrise."

Another pause.

"Did you get the berkelium?" Theodor finally asked.

Cassie fought her anger. Theodor wasn't being heartless. The team below was secondary to the thousands on the ship.

"Just a core sample, sir," she said. "But it confirms that the deposit's there."

Theodor swore. "Dr. Morant, our orbit decays in under twenty hours. Report immediately after docking to brief the command team." Theodor cut the link.

Cassie stared at the huge wormship, suddenly hating it, hating its strangeness. *Humans would never build something like that*, she thought.

Consisting of hundreds of torus rings strung along a central axis like donuts on a stick, the ship resembled a giant metallic worm. A dozen rings near the middle were slowly rotating, providing the few inhabited sections with an artificial gravity. The thousands of humans on the ship barely filled a fraction of it.

This wasn't meant for us, she thought. We shouldn't be here.

Humans had just begun to explore their solar system, when Max Bremer and his crew had found the wormships, three of them, outside the orbit of Pluto.

Abandoned? Lost? Or left to be found?

Found by the ever curious, barely-out-of-the-trees man-apes of Earth. Found with charted wormholes in Sol System. Found with still-only-partly-translated, we-think-this-button-does-this libraries and databases, and we-can't-fix-it-so-it-better-never-break technology. Incredibly ancient yet perfectly functioning Wormer technology.

Wormers. The inevitable name given to Earth's unknown alien benefactors.

Five years later, humanity was here, exploring the stars, riding like toddlers on the shoulders of the Wormers.

But Cassie no longer wanted to be here. She wished she were back on Earth, safely cocooned in her apartment, with Vivaldi playing, lost in one of her jigsaw puzzles.

She shifted uncomfortably in the hopper seat. Like every Wormer chair, like the ship itself, it almost fit a human. But not quite.

It was like forcing a piece to fit in a jigsaw--it was always a cheat, and in the end, the picture was wrong. Humans didn't belong here. They had forced themselves into a place in the universe where they didn't fit. *We cheated*, she thought, *and we've been caught. And now we're being punished.*

They faced a puzzle that threatened the entire ship. She'd had a chance to solve it on the planet.

And she'd failed.

Cassie hugged herself, trying to think. She was good at puzzles, but this one had a piece missing. She thought back over events since they'd arrived through the wormhole four days ago. The answer had to be there...

***

Four days ago, Cassie had sat in her quarters on the *Kepler*, hunched over a jigsaw puzzle covering her desk. The desk, like anything Wormer, favored unbroken flowing contours, the seat sweeping up to chair back wrapping around to desk surface.

Viewplates on the curved walls showed telescopic shots of Griphus. The walls and ceiling glowed softly.

Lieutenant David Stockard, Davey to Cassie, lay on her bunk watching her.

"Don't you get tired of jigsaws?" he asked.

She shrugged. "They relax me. It's my form of meditation. Besides, I'm doing my homework."

Davey rolled off the bunk. She watched him walk over, wondering again what had brought them together. If she could call what they had being "together"--sometimes friendship, sometimes romance, sometimes not-talking-to-each-other.

They seemed a case study in "opposites attract." She was a scientist, and Davey was military. She was dark, short and slim, while he was fair, tall and broad. She preferred spending her time quietly, reading, listening to classical music--and doing jigsaw puzzles. Davey always had to be active.

But the biggest difference lay in their attitudes to the Wormers. Davey fervently believed that the alien ships were meant to be found by humans, that the Universe wanted them to explore the stars.

To Cassie, the Universe wasn't telling them everything it knew. She felt that they didn't understand Wormer technology enough to be risking thousands of lives.

He looked at the puzzle. "Homework?"

"I printed a Mercator projection of topographic scans of Griphus onto plas-per, and the computer cut it into a jigsaw."

The puzzle showed the planet's two major continents, which Dr. Xu, head geologist and Cassie's supervisor, had dubbed Manus and Pugnus. *Hand and fist.* The western continent, Pugnus, resembled a clenched fist and forearm, punching across an ocean at Manus, which resembled an open hand, fingers and thumb curled ready to catch the fist. Colored dots, each numbered, speckled the map.

"What are the dots?" Davey asked.

"Our shopping list. Deposits of rare minerals. That is, if you believe Wormer archives and Wormer scanners--"

"Cassie, let's not start--"

"Davey, these ships are at least ten thousand years old--"

"With self-healing nanotech--" Davey replied.

"That we don't understand--"

"Cassie..." Davey sighed.

She glared, then folded her arms. "Fine."

Davey checked the time on his per-comm unit. "Speaking of homework, Trask wants surface team rescue procedures by oh-eight-hundred. Gotta go." He kissed Cassie and left.

Cassie bit back a comment that this was a scientific, not a military, expedition. The likely need for Trask's "procedures" was low in her opinion.

She would soon change her mind.

An hour later, Cassie was walking along the busy outer corridor of the ring segment assigned to the science team. Suddenly, the ship shuddered, throwing Cassie and others against each other.

The ship lurched again, and the light from the glowing walls blinked out. People screamed. Emergency lighting tracks along the base of the walls flickered on. After some hesitation, people started walking along the corridor again, but with nervous glances around and whispered exchanges.

Cassie began walking again as well, but something felt...different. No, something felt *wrong*. A few more steps confirmed it. With each step, she bounced a little higher. She swallowed as she realized what was happening.

The rotation of the ring was slowing. The ship, or at least this ring segment, was losing artificial gravity.

The ring continued to slow until she was nearly floating. Grabbing a handhold, she pulled herself to the closest wall as others did the same. There, in the darkened corridor, she waited, surrounded by whispers, shouts, and sobbing.

After about thirty minutes, the walls began to brighten. Cassie felt gravity returning like an invisible hand tugging at her guts, followed by a heaviness in her limbs. As full lighting returned and gravity approached normal, she started walking again, but on now shaky legs. Most people stood dazed, as if afraid to move, looking like scattered pieces in a jigsaw that before had been a coherent picture of normality.

What had happened?

The intercom broke through the rising babble of conversations. "The following personnel report immediately to port six, segment beta four for surface team detail." Twenty names followed. One was Davey's.

One was hers. What was going on?

An hour later, her questions still unanswered, she and nineteen others sat in a hopper as it left the *Kepler*. Hoppers were smaller Wormer craft used for ship-to-surface trips and exploration. With a tubular hull, a spherical cockpit at the head, and six jointed legs allowing them to rest level on any terrain, they resembled grasshoppers.

The team faced each other in two rows of seats in the main cabin. Cassie only knew two others besides Davey. Manfred Mubuto, balding, dark and round, was their xeno-anthropologist. Liz Branson, with features as sharp as her sarcasm, was their linguist. Four were marines. But the rest, over half the team, were mining techs. Why?

Davey addressed them. She'd never seen him so serious.

"The Kepler's power loss resulted from the primary fuel cell being purged. Engineering is working to swap cells, but that requires translating untested Wormer procedures. We may need to replenish the cell, which means extracting berkelium from Griphus for processing."

That's why I'm here, Cassie thought. Berkelium, a rare trans-uranium element, was the favored Wormer energy source. It had never been found on Earth. Humans had managed to manufacture it, but only in trace amounts and not the specific isotope needed to power a Wormer ship. Her analysis of Griphus had shown possible deposits of the necessary isotope.

"Like every planet found via the wormholes," Davey said, "Griphus is incredibly Earth-like: atmosphere, gravity, humanoid populations--"

Liz interrupted. "We purged a fuel cell? Who screwed up?"

Davey reddened. "That's not relevant--"

"Operator error, I hear," Manfred said. "A tech misread Wormer symbols on a panel, punched an incorrect sequence--"

Liz swore. "I knew it! We're like kids trying to fly Daddy's flitter--"

Cassie started to agree, but Davey cut them off.

"We've no time for rumors," he snapped, looking at Cassie, Liz, and Manfred. "Our orbit decays in three days. I remind you that this team's under my command--including science personnel."

Manfred nodded. Liz glared, but said nothing.

Davey tapped the computer pad on his seat. A holo of Griphus appeared. "Dr. Morant, please locate the berkelium."

Cassie almost laughed at being called "Dr. Morant" by Davey, but then she caught his look. She tapped some keys, and two red dots blinked onto the holo, one in the ocean mid-way between Pugnus and Manus, and another offshore of Manus. The second site was circled.

"Wormer sensors show two sites. I've circled my recommendation," Cassie said.

"Why not the other site?" a mining tech asked.

A network of lines appeared, making the planet's surface look like a huge jigsaw puzzle.

"As on Earth," Cassie said, "the lithosphere or planetary crust of Griphus is broken into tectonic plates, irregular sections ranging from maybe fifteen kilometers thick under oceans to a hundred under continents. This shows the plate pattern on Griphus.

"Plates float on the denser, semi-molten asthenosphere, the upper part of the mantle. At 'transform' boundaries, they slide along each other, as in the San Andreas Fault on Earth. At 'convergent' boundaries, they collide, forming mountains such as the Himalayas."

A line splitting the ocean between Pugnus and Manus glowed yellow. The line also ran through the other berkelium site.

"But at 'divergent' boundaries," Cassie continued, "such as this mid-oceanic mountain range, magma pushes up from the mantle, creating new crust, forcing the plates apart. The other site is located in these mountains, but in a valley between two peaks and below our mining submersible's crush depth."

Davey nodded. "So we hit the site offshore of Manus. Any indigenous population along that coast?"

"Yes," Manfred said. "From orbital pictures, they appear tribal, agrarian, definitely pre-industrial. Some large stone structures and primitive metallurgy."

"Then defending ourselves shouldn't be a problem." Davey patted the *stinger* on his belt. The Wormer weapon was non-lethal, temporarily disrupting voluntary muscular control.

"Could we try talking before we shoot them?" Liz said.

Davey just smiled. "Which brings us to communication, Dr. Branson."

Liz sighed. "Wormer translator units need a critical mass of vocabulary, syntax, and context samples to learn a language. Given the time we have, I doubt they'll help much."

"With any luck, we won't need them," Davey said. "We'll locate the deposit, send in the mining submersible, and be out before they know we're there."

Looking around her, Cassie guessed that no one felt lucky.

The hopper landed on the coast near the offshore deposit. The team wore light body suits and breathing masks to prevent ingesting anything alien to human immune systems.

Cassie stepped onto a broad beach of gray sand lapped by an ocean too green for Earth, under a sky a touch too blue. The beach ran up to a forest of trees whose black trunks rose twenty meters into the air. Long silver leaves studded each trunk, glinting like sword blades in the sun. She heard a high keening that might have been birds or wind in the strange trees.

Southwards, the beach ran into the distance. But to the north, it ended at a cliff rising up to a low mesa. Cassie walked over to Davey, who was overseeing the marines unloading the submersible and drilling equipment.

"Cool, eh?" he said, looking around them.

She pointed at the mesa. "That's cooler to a rock nut."

He looked up the beach. "Okay. But keep your per-comm on."

Cassie nodded and set out. The cliff was an hour's walk. Cassie didn't mind, enjoying the exercise and strange surroundings. She took pictures of the rock strata and climbed to get samples at different levels. Then she walked back.

They captured Cassie just as she was wondering why the hopper seemed deserted. The natives appeared so quickly and silently, they seemed to rise from the sand. Cassie counted about forty of them, all remarkably human-like, but taller, with larger eyes, longer noses, and greenish skin. All were male, bare-chested, wearing skirts woven from sword-blade tree leaves, and leather sandals.

They led Cassie to stand before two women. One was dressed as the men were, but with a headdress of a coppery metal. The other was older and wore a cape of cloth and feathers. Her head was bare, her hair long and white. Beside them, pale but unharmed, stood Liz Branson, flanked by two warriors.

The older woman spoke to Liz in a sing-song melodic language. Cassie saw that the linguist wore a translator earplug. Liz sat down, motioning Cassie to do the same. The male warriors sat circling them. The two native women remained standing.

Cassie realized she was trembling. “What happened?”

Liz grimaced. “We’ve stepped in it big time. The Chadorans--our captors--believe a sacred object called “the third one” lies underwater here. Only a priestess may enter these waters. When our techs launched the sub, the natives ambushed us from the trees with blowguns. They grabbed the techs when they surfaced.”

“Where’s Davey?” Cassie asked, then added, “...and everyone?”

“Taken somewhere. They seemed okay.”

“Why not you, too?”

“The tribe’s matriarchal,” Liz said. “The old woman is Cha-kay, their chief. The younger one, Pre-nah, is their priestess. Because I’m female and knew their language, Cha-kay assumed I was our leader. But I said you were.”

“You what?” Cassie cried.

“Cassie, we need someone they’ll respect,” Liz said, her face grim. “That means a female who didn’t defile the site. That means you.”

“God, Liz--wait, how can you talk to them?”

Liz frowned. “It’s weird. The translator produced understandable versions within minutes, pulling from Wormer archives of other worlds. That implies all those languages share the same roots. The Wormers may have seeded all these worlds.”

Cassie didn’t care. “What can I do?”

“Convince Cha-kay to let us go.”

“How?” Cassie asked.

“She wants to show you something. It’s some sort of test.”

“And if I fail?”

Liz handed Cassie the translator. “Then they’ll kill us.”

Cassie swallowed. “I won’t let that happen.”

They led Cassie to a long boat with a curving prow powered by a dozen rowers. Cha-kay rode in a chair near the stern, Cassie at her feet. Pre-nah and six warriors stood beside them.

They traveled up a winding river through dense jungle. Conversation was sparse, but sufficient to convince Cassie that the translator unit worked. After three hours, they landed at a clearing. Cassie climbed out, happy to move and stretch. She blinked.

Blue cubes, ranging from one to ten meters high, filled the clearing. They were hewn from stone and painted. The party walked past the cubes to a path that switch-backed up a low mountain. They began to climb.

Cassie groaned but said nothing, since the aged Cha-kay didn't seem bothered by the climb. As they went, Cassie noticed smaller cubes beside the path.

Night had fallen when they reached the top and stepped onto a tabletop of rock about eighty meters across. Cassie gasped.

A huge cube, at least fifty meters on each side nearly filled the plateau. It was blue. It was glowing.

And it was hovering a meter off the ground.

Cha-kay led Cassie to it, and Cassie received another shock. On its smooth sides, Cassie saw familiar symbols.

The artifact, whatever its purpose, was Wormer.

Cha-kay prostrated herself, telling Cassie to do the same. As Cassie did so, she peeked underneath the cube. A column of pulsating blue light shone from a crevice to touch the base of the artifact at its center. Reaching down to her belt, Cassie activated her scanner. She'd check the readings later.

Rising, Cha-kay indicated a large diagram on the artifact. In it, a cube, a sphere, and a tetrahedron formed points of an equilateral triangle.

"It is a map. We are here," Cha-kay said, pointing to the cube. "The gods left three artifacts, but hid one. The third will appear when the gods return and lay their hands on the other two." Then, pointing to the outline of a hand on the artifact, Cha-kay looked at Cassie.

"Touch," she said.

With a sudden chill, Cassie understood. *They think we're the Wormers, finally returning*, she thought.

This was the test, on which the lives of her shipmates, of the entire ship, depended.

Reaching out a trembling hand, Cassie felt resistance from some invisible barrier and a warm tingling, then her hand slipped through onto the outline on the artifact.

Nothing happened.

Murmurs grew behind her. Feeling sick, Cassie looked at Cha-kay. To her surprise, the old woman smiled.

"Perhaps," Cha-kay said, "it rises even now."

Cassie understood. Cha-kay hoped to find that the third artifact had emerged from the sea when they returned to the beach. Cassie didn't share her hope.

They spent the night there. Pretending to sleep, Cassie checked her scanner readings. They confirmed her suspicions. The column of light showed berkelium emissions. The artifact was connected to a deposit as an energy source.

The next day, a similar journey brought them to the second artifact, located on another flat mountain peak. The only difference was the artifact itself, a huge glowing red tetrahedron. Cassie again saw a column of light underneath and detected berkelium. She touched the artifact, again with no apparent effect, and the party began the trip back.

Cha-kay seemed to have grown genuinely fond of Cassie. She told Cassie how her people found the artifacts generations ago, eventually realizing that the drawing was a map. They learned to measure distances and angles, and determined that the third artifact lay in the coastal waters. Priestesses had dived there for centuries but found nothing. Still they believed.

Cassie did some calculations, and found the Chadoran estimate remarkably accurate. Still, she wondered why the Wormers would locate two artifacts in identical settings on mountain plateaus, yet place the third underwater. Perhaps the third location had subsided over the years. But her scans showed no sunken mountains off the coast.

Cassie enjoyed Cha-kay's company, but as they neared the coast, her fear grew. Cha-kay fell silent as well. As the boat reached the beach, they stood at the railing, clasping each other's hand, scanning the waters for the third artifact.

Nothing.

Cries arose among the warriors. Pre-nah approached Cha-kay. "The strangers are false gods," the priestess said. "They must die."

Cha-kay stared across the ocean. Finally, she nodded. Cassie's legs grew weak as two warriors moved toward her.

Cha-kay raised her hand. "No. This one goes free. She did not defile the sacred place."

Pre-nah didn't look pleased, but she bowed her head.

They landed, and Cha-kay walked with Cassie to the hopper.

"When?" Cassie asked, her voice breaking.

"At sunrise, child," Cha-kay said. "I am sorry."

Cassie boarded the hopper. She engaged the auto-launch, then slumped in her seat as the planet and her hopes grew smaller.

***

After docking, Cassie went immediately to the briefing room, as Captain Theodor had ordered. She quickly took a seat in one of a dozen Wormer chairs around a holo display unit. Dr. Xu gave her a worried smile. Commander Trask glared.

Theodor cleared his throat, a rumble that brought everyone's gaze to his stocky form.

"I'll be brief. Our orbit collapses in nineteen hours. Attempts to swap fuel cells were unsuccessful. The team sent to extract the berkelium has been captured and faces execution. Only Dr. Morant escaped."

Everyone looked at Cassie. All she could think of was how she'd failed.

Theodor continued. "Dr. Morant will summarize events on the planet. Then I need ideas."

Cassie told her story, then answered questions, mostly dealing with the artifacts. Will Epps, their expert on Wormer texts and writing, after analyzing her scans, agreed that the artifacts were Wormer.

The team began reviewing and discarding proposals. Finally, Theodor made his decision. A platoon of marines would drop outside the Chadoran city. Three squads would act as a diversion, drawing warriors from the city, while one squad slipped in for a search and rescue. One hour later, a hopper would drop two mining subs at the berkelium site.

“Sir, the priestess dives there daily,” Cassie said. “When they see our subs, they’ll kill the team.”

“That’s why I’m giving the rescue squads an hour head start,” Theodor replied. “It’s not much, but our priority is to replenish our fuel before our orbit decays. I can’t delay the berkelium extraction any longer.”

Cassie slumped in her seat. Davey, Liz, the others. They were all going to die.

Trask stood. “If Dr. Morant could provide a topographical display of the area, I’ll outline the attack plan.”

Cassie tapped some keys, and the planetary view of Griphus appeared, including the pattern of tectonic plates.

Like a jigsaw puzzle, Cassie thought. Why can't this be that simple?

“Zoom in to the landing site,” Trask said.

Freezing the rotation over Pugnus and Manus, Cassie started to zoom in, then stopped, staring at the display. *No*, she thought, *it’s too wild. But maybe...* She began tapping furiously, and calculations streamed across the holo.

“What the hell’s going on?” Trask asked.

Theodor frowned. “Dr. Morant?”

Cassie looked at her results. My god, it fits. But the time span...

“Dr. Morant!” Theodor barked.

Cassie’s head jerked up. Everyone was staring. *It’s wild*, she thought, *but it fits*. And she liked things that fit.

"Captain," Cassie said, "what if we proved to the Chadorans that the deposit site is *not* sacred?"

Theodor frowned. "Discredit their religion? I don't--"

"No," Cassie said. "I mean, prove that it isn't sacred because..." She stopped. *What if she was wrong?* But it was Davey and the team's only chance.

"...because the third artifact isn't there," she finished.

Trask snorted. "Then why will they kill to protect the site?"

"Because they *think* it's there, based entirely on the diagrams on the artifacts."

"And you think those diagrams are wrong?" Theodor asked, but his voice held none of Trask's derision.

"I think they were correct once," she said. "But not anymore."

"So where's the artifact?" Theodor asked.

Cassie's hand trembled as she tapped more keys. Two green lights appeared inland on the western coast of Manus, followed by a red light just off the same coast, forming the triangular pattern diagrammed on the artifacts.

"The two green lights are the known artifacts. The red light is both the supposed underwater location of the third and our targeted berkelium site."

She swallowed. *Here goes*, she thought.

"And this, I believe, is the actual location of the third artifact." A third green light appeared.

Everyone started talking at once. Theodor silenced them with a wave of his hand. He stared at the display.

On the eastern coast of Pugnus, on a separate continent and an entire ocean away from the underwater site, blinked the third green light.

Theodor turned to Cassie. “Explain.”

“It involves tectonic plate theory--” she began.

“I know the theory. What’s the relevance?”

Cassie tapped a key. The submerged mid-oceanic mountain range between Pugnus and Manus glowed yellow.

“That submerged range is a ‘divergent’ boundary,” Cassie said, “where new crust is being formed, pushing Manus and Pugnus further apart every year. But that also means that sometime in the past, they looked like this.” The plates began to shift. The two large continents moved closer until the fist of Pugnus slipped into the open hand of Manus like a piece in a puzzle. Someone gasped, as the third green light on Pugnus aligned itself over the red light offshore of Manus.

Theodor nodded. “You’re saying the Wormers originally placed the three artifacts as the diagrams show, but the missing one moved relative to the other two as the continents separated.”

Xu shook his head. “Cassie...”

Cassie sighed. “I know. The time frame is...difficult to believe.”

“How old are the artifacts if your theory is true?” Theodor asked.

Xu answered. “At least as old as the core sample from the deposit site, which formed as the submerged range grew and the ocean floor started to spread. Cassie, what was the isotopic clock dating on the sample?”

Cassie hesitated. “Its age was thirty, uh...” She swallowed. “...million years.”

The eruption of exclamations made Cassie want to slink from the room. Theodor again waved for silence.

In desperation, Cassie turned to Will Epps. “We know that these ships are at least ten thousand years old. But couldn't they be much older?”

Several people squirmed. Their situation was bad enough without being reminded that they were relying on alien technology at least a hundred centuries old.

Will shrugged. “There’s so much self-healing nano-tech, we can’t estimate their age accurately.”

“So any Wormer technology could be much older as well, right?” Cassie asked.

“But thirty million years...” Xu shook his head, as did others. Cassie was losing them.

She turned to Theodor.

“Captain, it all fits. It explains why the Chadorans have never found the artifact. Why our sub didn’t see it. Why Wormers placed two artifacts on mountains, but supposedly put the third underwater. They didn't. They put it on land too.”

“Can't we scan for the artifact?” Trask said.

“The other two don’t show on scanners,” Epps said. “They’re shielded somehow.”

“So the third artifact *could* be where the Chadorans say it is,” Trask replied.

Cassie sat back, feeling defeated. Then something struck her.

"Both artifacts I saw are located over berkelium deposits, yet neither site appears on the mineral scans. The artifacts shield the berkelium too."

"So?" Theodor said.

"We detected berkelium at the underwater site. That means nothing's shielding it. The third artifact isn't there."

Trask started to protest, but Theodor raised a hand. "I agree with Dr. Morant. It fits." He stood up. "Cassie, I'll give you the same lead time. Take a hopper down now."

Cassie was already sprinting for the door.

***

On a mountain plateau, across an ocean from where they had first landed on Griphus, Cassie and Davey stood, arms around each other's waist.

"So you saved me, the team, the entire ship," Davey said, "and made one of the most important discoveries in history. Not a bad day."

Cassie grinned. "Actually, the toughest part was convincing Cha-kay to fly in the hopper. Now she wants a world tour."

Beside them, happiness lighting her face, Cha-kay gazed at a huge glowing yellow sphere hovering above the ground.

The third artifact.

With one difference. A beam of energy shone from the sphere into the sky. The beam had begun the moment Cassie had touched the sphere.

Cassie's per-comm beeped. It was Theodor. “Dr. Morant, all three artifacts now appear on scanners, all beaming to the same point in space--”

“A new wormhole,” Cassie interrupted.

Pause. “How’d you know?” Theodor asked.

Cassie grinned. “I’m good at puzzles, sir.”

“Hmm. Anyway, Earth’s sending a second wormship. We’ll all have the option of returning home or exploring the wormhole. Once again, good work, Morant.” Theodor signed off.

“You didn’t mention your theory,” Davey said.

“That the wormhole leads to the Wormers’ home world? Just a hunch.”

“Explain it to me then.”

Cassie nodded at the sphere. “I think the artifacts were a puzzle--and the wormhole the prize.”

“For us or the Chandorans?”

“For us. Another bread crumb in the trail the Wormers left us.” She shrugged and laughed. “It just fits.”

Davey nodded. “So what about you? Back to Earth or through the wormhole?”

“Wormhole,” she said.

He raised an eyebrow. “Okay, that surprised me.”

Cassie grinned. “Hey, if the Wormers liked puzzles, they couldn’t have been that bad.” She stared at the artifact. “Besides, we solved their puzzle, saved ourselves, became heroes to the Chadorans..." Her eyes followed the beam up towards the heavens.

"Maybe we fit out here after all," she said softly.

***

Douglas Smith is a 5-time award-winning author described by *Library Journal* as "one of Canada's most original writers of speculative fiction." His work has appeared in 28 languages and includes the multi-award-winning urban fantasy trilogy, *The Dream Rider Saga*; the urban fantasy novel, *The Wolf at the End of the World*; the collections *Chimerascope*, *Impossibilia*, and *Borderlanz*; and the writer's guides, *Playing the Short Game: How to Market & Sell Short Fiction* and *Brick by Brick: How to Build a Story*.

.

Art Credit: Jigsaw artwork by Erin Colson.

"Jigsaw" first appeared in Odyssey (Fitzhenry & Whiteside Press, Canada; 2004; Julie Czerneda, ed.). It was a finalist for the Aurora Award in Canada in 2005. It was reprinted online in *Issues in Earth Science*, Issue 4, July 2015.

Chase Your Own Rabbits!

Question: Find something in the story that you wonder about, or want to challenge, or would like to explore more fully, and ask a scientifically-investigable question about it. As you consider things you wonder about, remember that not all questions can be investigated scientifically. For example, in the story *Jigsaw*, experimentally measuring the claims or interpretations is hampered by the vast time scales involved, and so any experiment needs to be constrained to a small piece of the puzzle. A question like "What is the radioactive half life of berkelium" can involve experimental investigation, but it would be beyond the scope of the classroom because 1) berkelium does not occur naturally and must be manufactured, 2) it is dangerously radioactive and 3) the isotope in the story is imaginary and does not exist in the known real world. Try to find a question that can be investigated scientifically with resources that you have, or can gather, in the classroom.

The question we, Russ and Mary, address in our Learn by Example for the story *Jigsaw* is the following:

"How can we use observational evidence combined with modeling to place constraints on the events that happen in the story"

You can choose this question as a seed-question for your own investigation or come up with your own rabbit to chase!

Feel free to write down your question here:

Vague Notion: As you prepare to plan your investigation, first get a vague notion of what you plan to do—including guidelines on whether it will involve experimentation, data-interpretation, modeling, arguing from evidence or some other practice of science.

The vague notion that we started from for our Learn by Example for this story is the following:

With an understanding of the model for plate tectonics, can we figure out more about the Griphus world and where the third deposit of shielded berkelium must be located? Based on mathematical modeling of radioactive decay, can we place constraints on the mysterious isotope of berkelium that plays such a significant role in the story?

If you chose to go with our question for your own investigation, you can also choose our starting vague notion, or chase your own rabbit.

Feel free to write down your vague notion here:

Detailed Planning: Once you have your question and a vague notion in place, you can begin to fill in the nuances, complexities, and details of your plan. There is no recipe for your plan or a predetermined set of steps to follow, because your plan ***is*** the recipe or set of steps that you are creating for this particular investigation! Sketch out your investigation, anticipating as best you can what challenges you will encounter, and include as much detail as you are able (realizing that your investigative plan may not survive the first results of actually doing it.) Be sure to think about what background information you might need and how to get it, techniques you may need to research, the overall sequence of things you will need to do, and all other aspects of your investigation! You will likely need several pages to document your plan.

Follow a Question

Asking a question, choosing a vague notion of how to pursue that question, and then developing a detailed plan to pursue it is just the start of investigating. You actually have to do your investigation! During the course of doing your investigation, you will find out what works and what doesn't work. You may realize that you need more information, or different experimental tools, or even that your initial working model was so far off that your plan for investigation quickly falls apart. Authentic investigation involves iterative thinking during which your original experimental plan is revised, your initial interpretations updated, and your working model revisited. All of this is the whole point and practice of authentic investigation!

Keep careful notes on your experimental methods, interpretations, and models as well as your revisions as you work through your investigation.

Learn by Example (for Jigsaw)

Our Starting Question: "How can we use observational evidence combined with modeling to place constraints on the events that happen in the Jigsaw?"

Our Starting Notion: With an understanding of the model for plate tectonics, can we figure out more about the Griphus world and where the third deposit of shielded berkelium must be located? Based on mathematical modeling of radioactive decay, can we place constraints on the mysterious isotope of berkelium that plays such a significant role in the story?

Introduction

An understanding of plate tectonics is an essential element to the plot of this story, a finalist for the Canadian Aurora award, one of the most prestigious science fiction awards in Canada. As science investigators, we should be delighted that real science investigation found its way into such a prominent place in literature! As geologists, can we follow the logic of Cassie Morant, and perhaps even expand on it? For this investigation, we presume a preexisting understanding of plate tectonics, including types of boundaries and associated features.

Remember, the following puzzles and challenges are not intended to be systematic curricular steps through which students of science work toward defined learning outcomes. Rather, they are intended as aspirational examples of investigative thinking, appropriate at a variety of levels—some middle school, some high school, some for the expert teacher—to inspire investigative thoughts and increase understanding of what investigative thinking is all about. Find one or more that is right for you!

Cassie figured out the location of the third shielded deposit of berkelium, which should have been located on the third apex of a triangle connecting the three geometric artifacts. The Chadorans thought the artifact was located at that apex, off the coast of the continent of Manus, the hand. The Earthers' initial discovery of a berkelium deposit at this sacred spot is the story's inciting event.

She swallowed. Here goes, she thought. "And this, I believe, is the actual location of the third artifact." A third green light appeared. Everyone started talking at once. Theodor silenced them with a wave of his hand. He stared at the display. On the eastern coast of Pugnus, on a separate continent and an entire ocean away from the underwater site, blinked the third green light.

Below is a map that we, Russ and Mary, constructed based on the story, showing the location of the first two artifacts, the cube (blue square) and the tetrahedron (red triangle), as well as the presumed location of the third artifact, the sphere (purple circle), at the apex of the yellow triangle.

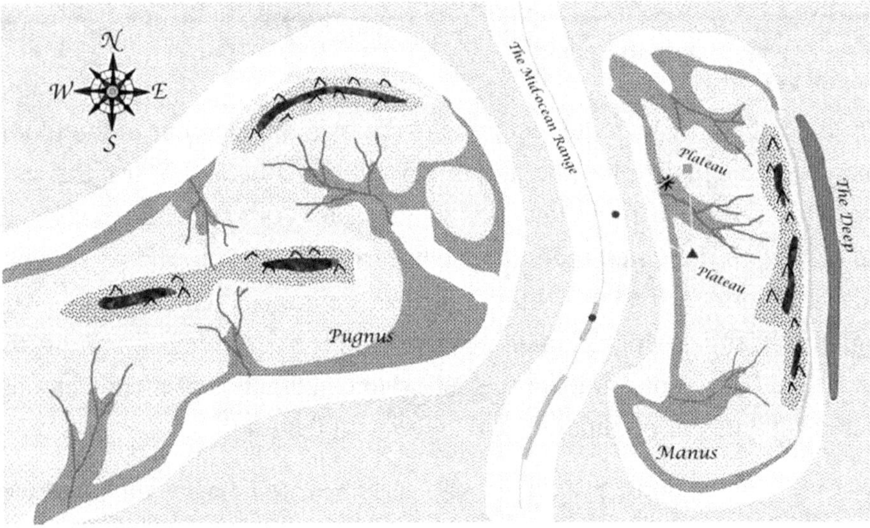

Challenge 1: The Third Light

Where do you think Cassie's third green light appeared, marking the location of the spherical artifact? Why do you think that?

Spoiler Alert: Mentally moving Pugnus to fit into the shape of Manus, and taking into account the distance that the continents have moved apart by plate tectonic motion, the location might be somewhere in the circle shown on the map provided further below. We determined this location by measuring the distance from the middle of the Manus continental shelf to the purple dot at the apex of the yellow triangle, and then measuring that same distance inland from the middle of the Pugnus continental shelf (this method overlaps the continental shelves by about half, which we intended to approximate extension during rifting, although skipping this overlap by measuring to the edge of the continental shelf in each case might also be a valid approach, yielding an essentially-identical location).

Challenge 2: Interpreting Plate Boundaries

Based on the story, and on the map of Griphus, identify the location of any of the following types of plate tectonic boundaries: 1) continent-continent convergence, 2) continent-ocean convergence with subduction, 3) ocean-ocean divergence, 4) continent-continent divergence. Cite the evidence seen on the map for each that makes you think it might be present.

Spoiler Alert: The locations marked by the letters A, B, C, D, and E on the map below might be current or former plate boundaries.

A. Possible convergent boundary between continental and oceanic crust. Evidence is the presence of a mountain range on a coast. However, there is no trench, suggesting that this might be a remnant mountain range from past convergence that is no longer active.

B. Likely convergent boundary between continental and continental crust, which might still be active. Evidence is the presence of mountains with continental crust on either side.

C. Possible continent-continent divergence. Evidence is the somewhat straightish appearance of what could be a rift valley with continental crust on either side. However, this could also be an erosional valley feature which later flooded with a rise in sea level. Additional information, such as whether there had been any volcanism associated with the feature, would be needed to confirm.

D. Likely active continental-oceanic convergence. Evidence includes the coastal mountains and the trench running parallel to them offshore.

E. Likely active oceanic-oceanic divergence. Evidence is the long, sinuous underwater mountain range seen by the presence of the shallower depths of the ocean and the symmetrical, centered relationship of this sinuous ridge relative to the coastlines on either side of the ocean.

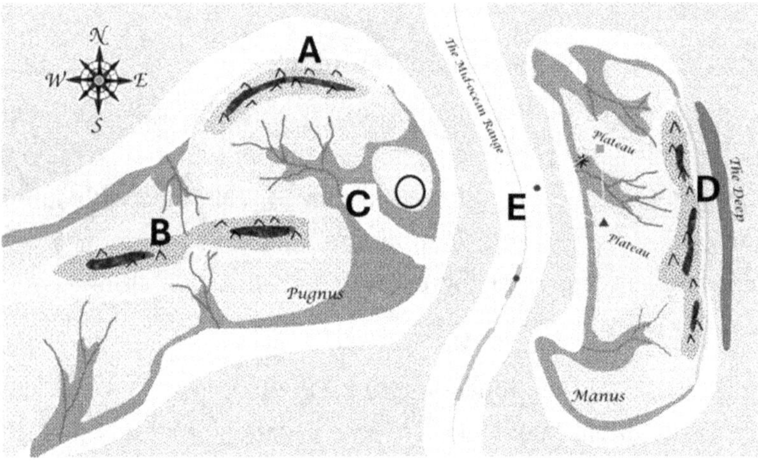

Challenge 3: Age of Ocean Crust

Based on the story, illustrate the ages for the ocean crust between Manus and Pugnus. Include both the pattern and value of the rock ages across the ocean between the two continents.

Spoiler Alert: The ages of rock should increase outward in a symmetrical pattern from the center of the mid-ocean range, such the rock in the middle of the ocean is youngest, getting progressively older outward in a series of belts that roughly parallel the range. The age in the middle should be 0 years old (forming today) and the age on the farthest sides of the ocean (what formed when Pugnus and Manus started rifting apart) 30 million years old, as per the ages given in the story.

Challenge 4: How Wide is the Sea?

The story doesn't really tell us the width of the ocean between Pugnus and Manus, growing ever wider over the 30 million years it has been forming. However, if rifting rates are similar to those on Earth, we can estimate how big the ocean might be. Use the map below (modified from the map of ocean crustal ages created by Mr. Elliot Lim, 2008, CIRES & NOAA/NCEI) to estimate how much the Atlantic Ocean widened per million years on Earth, and then use that number to estimate how wide the ocean on Griphus might be.

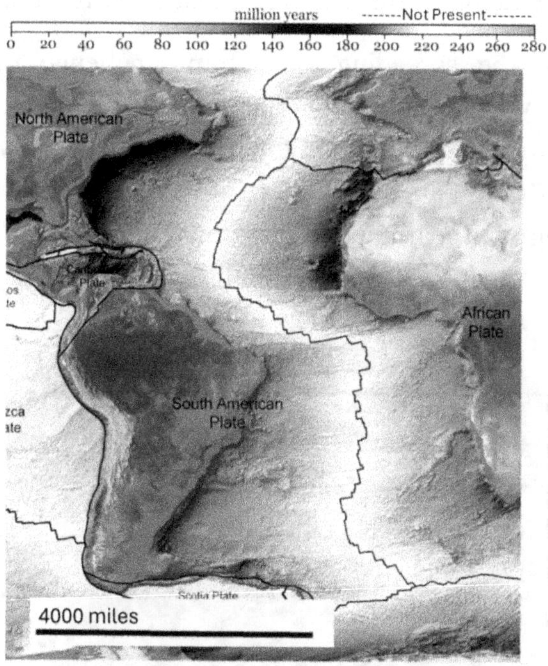

Spoiler Alert: Based on the map, the oldest rocks along the east coast of North America and the northwest coast of Africa might be 170-180 million years old. The distance across the ocean at this point is a bit less than 4000 miles. Taking 3800 miles, roughly, and dividing by the number of years, gives us 3800/175 million = 21.7 miles per million years. So, on Griphus, the ocean might be 21.7 miles per million years x 30 million years = 651 miles wide (not too big an ocean yet!)

Challenge 5: Testing our Measurements

Sometimes we might wonder "wait, what if our measurements of the age of rocks are wrong—how can we really know the rate of spreading?" It's always a good idea to test our scientific conclusion against other data that does not depend on the same measurements to see if they "match up." For example, we can use GPS measurements to determine the present-day rate of plate movement and compare it to our estimates based on the ages of rock in the Atlantic Ocean. In this challenge, we use spreading rates measured in modern times with GPS along the Atlantic Rift, to see if the rates match up with our estimate from the past 175 million years.

First, calculate the spreading rate from Challenge 4 in terms of centimeters per year (a bit of conversion math), and then compare to the measurements of spreading rate at the rift in Iceland shown in the graph below (adapted from data presented in "Crustal deformation in Iceland: Plate spreading and earthquake deformation" by Thóra Árnadóttir, Halldór Geirsson and Weiping Jiang published in JÖKULL No. 58, 2008.)

What do you think? Be sure to read the graph labels!

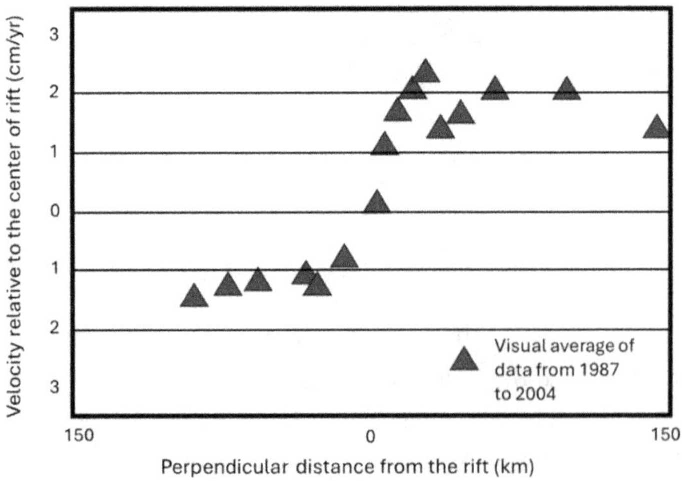

Spoiler Alert: Converting 21.7 miles per million years to centimeters per year:

21.7 mi/my x 1.61 km/mi x 100,000 cm/km ÷ 1,000,000 yrs/my = 3.5 cm/yr

The value of 3.5 cm/yr corresponds quite well to the sum of the velocities of the two sides of the Iceland rift (the spreading rate) of perhaps 3cm/year or a bit over, suggesting that 1) the rate of spreading may vary somewhat over time, but is close today to the average over the past 180 million years, and 2) our measurements for the age of the rocks on the floor of the Atlantic ocean are probably not far off.

Challenge 6: Radioactive Half-Life—A Thought Experiment

The Wormers—and now our heroes—used berkelium for power. In real life, no isotope of berkelium occurs naturally on Earth, although several isotopes have been manufactured artificially. It is an interesting math exercise, and an interesting exercise in understanding the idea of half-life, to consider why the isotope of berkelium in the story *cannot* be one of the isotopes that have been manufactured artificially on Earth, meaning that the berkelium isotope in the story must be purely fictional.

What we know: The longest half-life for any known isotope of berkelium is about 1380 years. This means that after 1380 years, half of an original amount of berkelium will have decayed to form daughter elements. The planet Griphus must be at least 30 million years old to fit with the story (and would reasonably be much older).

PUZZLE 1: How many atoms of berkelium would be left after 30 million years?

Let's figure this out in a series of steps.

STEP 1: How many half-lives pass in 30 million years? (Spoiler Alert: 30 million divided by 1380).

STEP 2: Suppose that Griphus started out as pure berkelium (an absurd conjecture...but let's suppose). Perhaps about 6×10^{49} atoms of berkelium.

That's 60 thousand billion billion billion billion billions. Quite a lot.

How many would be left after 1380 years? (Spoiler Alert: half, or 3×10^{49} atoms of berkelium)

STEP 3: How many after 2 x 1380 years = 2760 years? (Spoiler Alert: ½ x ½ = one-fourth = 1.5×10^{49} atoms of berkelium).

STEP 4: How many would be left after 207,000 years (150 half-lives)

150 half-lives corresponds to 150 halvings. Or $0.5^{150} = 7\text{x}10^{-46}$.

(Spoiler Alert: On average, there would be only about $6\text{x}10^{49} \times 7\text{x}10^{-46}$ atoms of berkelium = 42,000 atoms left.)

How many in another 1380 years? (Spoiler Alert: half, or 21000 atoms)

How many would be left after 227,700 years (165 half-lives)?

(Spoiler Alert: 165 half-lives corresponds to 165 halvings. Or $0.5^{165} = 2.1 \text{ x}10^{-50}$. On average, there would only be a single atom of berkelium left after only 227,700 years = $6\text{x}10^{49} \times 2.1\text{x}10^{-50}$.

And the planet has at least 29,772,300 years to go.

PUZZLE 2: How long would the half-life of the imaginary berkelium isotope in *Jigsaw* need to be in order to have a reasonably large deposit left after 30 million years?

Spoiler alert: Let's suppose that we need to have between 5% and 10% left to be meaningful, say 6%. Six percent approximately equals 1/16 = 1/(2x2x2x2) or 4 half-lives. So, to have about 6% left after 30 million years, the half-life must be at least 30,000,000/4 = 7.5 million years. That would be quite long compared to the known half-life of berkelium-247 of 1380 years! So, the mystery berkelium isotope seems quite unlikely to really exist in our dimensional reality!

Reflection on Learning

The goal of this book is to practice investigation itself, not to focus on the information and ideas you understood better through investigation. If traditional learning outcomes are not the goal, then what have you learned? Taking some time to consider what you gained from your investigation, reflecting on the practice of science *metacognitively*, can help you hold onto what you have gained. For example: What questions did you ask? What observations and interpretations did you make? Did you *do* modeling, rather than just learning or applying a model?

Real investigation is fluid, creative, and defies easy categorization. Even so, it might be helpful to consider the following categories of investigation offered by the Next Generation Science Standard's science and engineering practices (NGSS Lead States. 2013. Next Generation Science Standards: National Academies Press).

- Asking questions and defining problems
- Developing and using models
- Planning and carrying out investigations
- Analyzing and interpreting data
- Using mathematics and computational thinking
- Constructing explanations and designing solutions
- Engaging in argument from evidence
- Obtaining, evaluating and communicating information

Notice that we have not numbered these practices. They are not steps in a sequential process, but rather components of an integrated and iterative process. Also recognize that this list of practices is **a** way to organize and understand the investigative practice of science, not **the** way. As you go through this reflection, be flexible in thinking about what aspects of your own investigation might be practices of science, because these practices can show up in unexpected ways. Remember, don't think so much about

what you learned or understood, but rather think about what you initiated, did, practiced or experienced.

On a separate sheet of paper, jot down some things that have strengthened your ability to pursue investigative thinking. Give examples of how you engaged in particular, specific practices.

Exploring with you the experience that we, the editors, had in developing our Learn by Example section might provide some guidance as you think about your own practices of science. For the preceding stories, we started with elements of our investigation and considered what practices those elements engaged us in. This time, let's reverse that and start with practices and ask what aspects of our investigation used them.

As before, we certainly engaged with some traditional learning of the information and theories of science. We started with a preexisting understanding of the theory of plate tectonics, but we reviewed features and characteristics of the different types of plate boundaries and also learned some new things about plate divergence rates, such as the direct measurements made in Iceland.

However, learning information or theories was not our goal. Our goal was to practice investigation. The practices of science that we highlight for this Learn by Example in the Table of Contents include modeling, interpretation, and mathematical thinking. In what way did we, the investigators who created and implemented this investigation, gain experience with these three practices? Here are a few examples.

To develop the map of Griphus, we had to use our knowledge of features that form at each of the different types of plate boundaries to create geographic constructions of the world that make plate tectonic sense. This is a type of **modeling**. We also **modeled**, mathematically, the width of the sea on Griphus and the half-life of the imaginary berkelium isotope.

To check our map for correctness, we had to **interpret** the observational features that we had drawn, to see if they were consistent with other aspects of Griphus and with current understanding of plate tectonics. We also had to compare the results of our calculations of rate of seafloor spreading based on crustal ages to those based on GPS measurements and consider if those values matched. This is also a type of **interpretation**.

We went through several mathematical calculations in this Learn by Example. However, a math calculation in itself does not constitute mathematical thinking. To develop these puzzles, we had to figure out *how* to calculate the width of the sea on Griphus. The **mathematical thinking** experience came not in doing the calculation, but in figuring out what observational data we needed to use and how to set up the calculation. Likewise, the **mathematical thinking** for the determination of the imaginary berkelium half-life is not found in figuring out how to do calculations with all those giant exponents (!), but rather in conceiving and implementing how to set up the problem in a logical way. The proportional thinking required in reading maps, and in comparing spreading rates on Earth and Griphus, is also **mathematical thinking**.

We also engaged with asking questions, constructing explanations, arguing from evidence, and other aspects of authentic investigation. To be authentic, your investigation need not engage in all the practices. However, real investigations often include many of these practices because the practices are not separate components but are parts of an integrated whole that we call scientific investigation.

Story by Jacqueline Howland

Dinosaur Dig

By Jacqueline Howland

Can a love of art inspire Jace to solve a geological puzzle, connect with his stepdad, and shut up the local bully all in one afternoon at the dig?

Sighing, Jace stared morosely out the passenger window of his stepfather's Jeep at the high desert scenery they were passing. Six weeks of looking for bits of old bone, because that's what his stepfather did. Six weeks of being light years from LA and his friends, and no electronic devices allowed, so he couldn't call or text or even listen to his music. At least he'd been allowed to bring his drawing stuff. He could see possibilities in the dramatic rise of lava tubules around them, thrusting into the deep blue sky like fingers pointing toward the light. Still, he'd so much rather have gone to Mexico with his mother. But his stepdad had said no, she'd be busy enough taking care of her dying grandmother along with his five-year-old sister. Half-sister, he'd wanted to correct. Your kid. He knew the score about that and this trip. He'd overheard his stepdad saying Jace shouldn't go to a place that held such painful memories.

Yeah, his real dad had gotten sick and died when they went to visit relatives in Mexico. Sure, Jace missed him and the way his dad, a climatologist, had so patiently explained the difficult processes he studied and anything else Jace didn't understand. But he'd been seven then, a long time ago! He couldn't tell them he'd listened in, though. They'd get mad. Instead, he'd tried to argue he could help his mom. "I'm not a little kid anymore, Dad. I'm sixteen. I watch Alyssa all the time. Besides, I could practice my Spanish. Colleges love for people to have a second language." His parents expected him to get a college degree. That was cool, but his stepdad wouldn't like his desire to go for art or music. Jace was expected to do science like his dad and stepdad, or even "soft" science like his mom, a psychologist. They wouldn't listen when he told them his brain didn't want to go that way. He shook his head a tiny bit at this thought. They just wouldn't listen to him.

"Grab your gear, Jace."

Jace started. He'd been lost in thought again and missed their arrival. He looked around at the other parked Jeeps and Broncos.

Mr. Renfro suddenly appeared, as if he'd been lurking behind one of the other cars just waiting for his boss to show up. "Good afternoon, Brian. I'm sure you'll be pleased with the progress we've made since you were last here." Mr. Renfro's pale face was polite under his broad brimmed hat, the second in command showing respect for his superior. Jace shuddered, knowing full well what—or rather *who*—always came with Mr. Renfro.

"Of course, James. As soon as we've dumped our gear, I'll visit the dig." Jace's stepdad smiled easily, the deeply tanned skin of his face lined beyond his forty-seven years. A big, blonde man, he talked easily with anyone, never shy or awkward. Jace, small, dark, and quiet, always felt eclipsed by the man. Jace could never feel as confident, especially when he was around so many super smart scientists. He just couldn't compete in this arena. Ironically, he understood quite a bit of what his stepfather did. He just couldn't seem to communicate that. His brain seemed to freeze whenever he felt like he was being put on the spot.

Jace had found the idea of digging up dinosaur bones cool once upon a time, but certain classes, especially math and science, had become increasingly hard for him. At one point, the term "learning disability" had been raised. He remembered his stepfather frowning and telling him he just had to work harder. That hadn't helped, but a nice older woman, a former teacher who tutored him, had made a difference. "You don't have a disability, Jace. You just see the world differently and you need a different approach to learning." She'd helped him with percentages and fractions by pointing out that he knew how to make change, so he already understood those things. And he'd had a substitute science teacher once who'd taught almost entirely with a hands-on approach. They'd done tons of experiments during the two weeks Mr. Hagar was there. Jace learned more during that time than the rest of the school year. Why couldn't all teachers be like that? He sighed. At least he was good at drawing and playing music.

They put their gear in the main tent and then went back out into the bright sunshine. For a moment Jace was blinded. Then he froze, his mouth hanging open. A girl stood there, clearly a part of the dig by her khaki shorts and shirt, but she could be an actress or something. She had skin the color of golden honey, auburn hair, bright eyes, and a ready smile. Not a girl. A young woman. He sighed.

With dancing eyes and the happiest smile Jace had ever seen, the young woman said joyfully, "Welcome back, Dr. Lamour. How are you?"

"Fine, fine, Lacy. How's your thesis coming?" By this, Jace knew that she was a graduate student. His parents had both talked about writing a thesis for their degrees. "Oh, this is my son, Jace." His stepdad turned to him as did Lacy. He felt himself blushing, but she probably couldn't tell. He just got a little darker.

"Hi, Jace. Nice to meet you!" She gave him a huge smile and all he could say was, "Uh."

Her smile faded a tiny bit, but she turned back to his stepfather, still sounding elated. "Well, we've got a bit of a mystery on our hands. You'll just never guess! Want to come see?" They walked away, chatting, and Jace stared after them.

A snide voice in his ear startled him. "Hey, bozo. She was really scintillated by that insightful comment of 'uh.' What are you going to follow that up with? Um? Er? Duh? Can't wait to hear your next foray into flirting!"

Jace turned and glared at Todd Renfro, older by a year and a lot taller. As usual when alone with Jace, he showed pure contempt. Jace and Todd had been pushed together often through the years since Todd's father worked for his stepfather, which Jace had always hated. Sure, Todd shone when it came to science, but he was also a jerk about it. He loved to make other people feel stupid. Just like with using that big word Jace had never heard before. He said nothing, just walked after his stepfather and Lacy. Better to be around more people. Todd was slightly less nasty then.

They reached the site where people were standing around, staring at the rocks. They seemed to see something important which was invisible to Jace. Mr. Renfro was explaining. "So, we've clearly got dinosaurs from the early Cretaceous here, a lot of them. You can see this excellent specimen of a Sauropelta here." Smiling, he pointed to darker bits, which Jace now noticed, encased in the bedrock at their feet. When he drew pictures, he had to pay close attention to those sorts of details. Maybe he should look at rocks as if he were going to draw them. "We've got unbroken bony plates of the head and even some unbroken neck spines plus more indicated. Possibly complete. A very exciting find."

"Hmm, yes, a nice find," his stepfather repeated, glancing at Lacy. "But we dug up that piece of Stegosaurus last year above this layer. That doesn't make sense. We haven't seen any evidence of crustal faulting or folding here." Something about his stepfather's expression made Jace wonder. He had that expectant look—a look that Jace had seen so much of, but always seemed to fail to fulfill. He guessed his stepdad did that to everyone.

Jace had a fair understanding of this stuff since his stepdad talked about all of it so much, but naturally Todd had to butt in. With mock kindness, he said, "We should explain for the benefit of Jace. He doesn't know that the Jurassic Period had

Stegosauruses and that it came before the Cretaceous Period. He probably can't even figure out that the material from the older period should be *under* the newer period, not on top of it. That's called the law of superposition. Unless there's crustal overturn, which can happen in a bunch of different ways like faulting or folding of the rocks. But he doesn't want to know those reasons. He doesn't want to know any of it. He hates science, don't you, Jace?"

Jace, barely noticing the disapproving looks on the adult's faces, gave Todd a venomous look, staying silent. He wasn't really that clueless about science even if he didn't know all the big words. It wasn't that hard to understand that the new stuff would be on top of the old stuff: anybody who threw clothes on the floor instead of putting them away knew that! But if he said anything, there'd be trouble.

Lacy smiled gently at Jace and said, "It's not so hard if you approach it in the right way. What *are* you interested in, Jace?"

He blushed, looked down, and answered softly, "Music. And art."

"Great!" she said. "Do you just listen to music or do you play too?"

"I play. A little." He looked at her and saw she was nodding.

"Then you know how music is written, with some notes above the line and some below. Think of a sheet of music as the earth's crust. Then imagine if you folded that paper just right, you could put those lower notes above the top notes. Or the paper could be folded multiple times, layering the crust or turning it into an accordion shape. And if you had a sheaf of different colored paper all folded the same way, it would be like the rock layers of the crust. You can see examples of these sometimes after excavations for roads where the grader slices into the side of a hill."

His face lit up. "Oh, yeah. I see." He had already got the folding thing, but it was so cool that she understood!

Lacy added, "And if you tore that paper lengthwise and then taped it back together so one half was partly raised and the other partly lowered, you'd get the equivalent of deformation from a fault line. Then you'll see where the lines of rock layers don't match on either side of the fault line." Her smiling enthusiasm grabbed Jace, who responded with excitement.

"Yeah! I see. And that's what all the stuff deep in the earth does, like how it creates earthquakes and volcanic eruptions." That had been part of Mr. Hagar's lessons. It came back to him easily now.

Todd spoke suddenly, with dripping sarcasm. "Gee, suddenly Jace likes science. I wonder why? Maybe because he's got a crush on Lacy?" Jace felt himself flushing. He was more aware, though, of a rising tide of rage.

He wanted so badly to say or even do something to Todd, but in his mind, he heard his mother's voice. "Watch your temper, Mijo. If you need a time-out to get control of yourself, then just say that." He turned, gritting his teeth, and walked away. Behind him he heard his stepfather sigh.

Jace walked past the camp and up a hill beyond it, wanting to get as far away as possible. Bad enough when it was just Todd making him look stupid and the others assuming he was. Now he was completely humiliated. Lacy could only despise him as a clueless kid. He trudged along, sliding on the rocky debris that covered the hillside until he found a rock free path. Water had probably cleared it. He knew about erosion. His dad had explained that on the trip to Mexico and a visit to some ancient ruins. He'd helped Jace understand how thousands and thousands of years of rain had broken down the seemingly indestructible rock building materials, enlightening Jace to the incredible power of natural forces. He had felt awed by the way water could push huge boulders, and how huge mountains were broken down by erosion over long periods of time.

Jace stared at the debris intently, thinking about drawing, always a soothing activity when he was upset. The rocks on the hillside looked a lot like the loose ones around the dig. Maybe water had carried them all the way down the hill and deposited them at the dig site. Jace looked ahead, noticing high walls of sheer rock rising straight up from the top of the hill. Why hadn't he seen that before? Maybe because he just thought of the places where his stepfather did his work as empty, boring areas. But the rock ahead of him was quite striking. He could draw it.

He stared, noting the details. Many cracks ran vertically so the wall wasn't solid. In fact, it looked unstable. Jace could see how columns had pulled away from the wall behind. For a moment, he felt some panic. What if one suddenly broke free and all that falling rock crushed him? A second later, he felt giddy. He thought he knew the explanation for the older bones being on top of the newer bones. He turned and raced back down the hill.

Rushing back to the dig, Jace felt elated by his revelation. What if there had been an unstable wall of rock like this back in the Cretaceous, a cliff made of what would have at that time been ancient Jurassic rock with stegosaurus bones in it? Those bones might have weathered out and fallen just like chunks of rock from this cliff were doing today. The bones from the Jurassic could have gotten mixed in with Sauropelta at that time and then been buried in sediment where they'd slowly turn into a new rock, a Cretaceous rock containing bones from both the Cretaceous and the Jurassic.

The workers had gone back to chipping away at the rocks. His father and Mr. Renfro stood talking nearby, with Todd hovering around them, trying to look important. Jace noticed Lacy working near them.

His stepdad gave him a concerned look as he approached but went on with what he was saying. Jace waited until he'd finished speaking before saying eagerly, "Dad. I know how that dinosaur you found last year wound up above newer bones!" His stepfather looked at him in shock, while Todd burst out laughing. Todd's dad gave him a stern look, though, and he shut up, but continued to smirk.

"OK, I'm listening, Jace." His stepfather wore his patient look.

Jace took a deep breath, feeling nervous. "OK, see those walls up there?" He pointed, and the others automatically looked. "I think those are that faulting thing. You know where some of the rock gets pushed up past the rest of the crust?"

His stepdad nodded and smiled faintly, amused maybe. But Jace pushed on anyway, determined to show them all he wasn't ignorant. "Well, the wall isn't solid. It's breaking down and chunks are falling off. That could have happened in the past, with another cliffside weathering away and then the chunks getting carried down by flooding. If the older bones were in those chunks, they could have wound up on top of newer bones, right?"

Lacy suddenly spoke excitedly. “Yes, I can see that. Wow, why didn’t I think of that? I forgot how important it is to look at the surroundings and not just what’s right in front of me.” She shook her head. “How did you figure that out, Jace?’

He shrugged, answering shyly, “Well, I used your explanation. It really helped me picture the stuff you were talking about. I was thinking about drawing the rock, so I really paid attention. I have to notice details when I draw, you see.”

Lacy smiled. “That’s cool. You know, you might think about becoming an illustrator for science textbooks and that sort of thing. You could combine your love of art with scientific knowledge.”

Jace straightened with a look of wonder. “Hey, that’s a good idea. It would be great to get paid for drawing.”

Jace noticed his stepfather grinning and suddenly realized he must have had the explanation all along. “This wasn’t a mystery for you at all, was it, Dad?”

His stepfather shook his head. “No. We knew what was going on. But it’s always better for people to find answers for themselves if they can. And I’ve always known you could do that if you’d just stick with things. Just because something is hard doesn’t mean you have to give up on it. We’ve all struggled with difficult scientific concepts. I think Lacy’s suggestion is a good one. You could use your artistic talents to help others understand science concepts. I think that would be a wonderful career for you, son.”

Jace nodded, excited by the idea and pleased by his stepfather’s approval. “Yeah, that’d be cool because I understand how it’s hard to see things sometimes. Now I know how to look. I just need to use my artist's eye and pay careful attention and maybe think about comparing what I don't know to something I do know. That really helps me." He smiled, noticing Todd in the background scowling. Maybe this trip would be fun, after all.

***

Jacqueline Howland is a retired college English instructor who enjoys gardening and bird watching. She also volunteers at her local science museum and has written blogs for them on various science topics including a recent one on fossils.

Art Credit: Dinosaur Dig artwork by Erin Colson.

"Dinosaur Dig" first appeared online in *Issues in Earth Science*, Issue 10, November 2018.

Chase Your Own Rabbits!

Question: Find something in the story that you wonder about, or want to challenge, or would like to explore more fully, and ask a scientifically-investigable question about it. As you consider things you wonder about, remember that not all questions can be investigated scientifically. For example, in the story *Dinosaur Dig*, it's possible that the researchers simply made a mistake in recording the layers where they found the mysteriously older fossils, but that isn't really possible to test through observation both because we can't reproduce field observations in the classroom and also because the field site is fictional and not real. A question like "Can I experimentally determine what kind of rock that the bones would be in if they were deposited in a river channel" could be addressed with classroom experiments with deposition and erosion using some kind of trough for a stream channel, although the kind of rock is not really considered in the story. Try to find a question that can be investigated scientifically with resources that you have, or can gather, in the classroom.

The question we, Russ and Mary, address in our Learn by Example for the story *Dinosaur Dig* is the following:

"Can we figure out the stratigraphy of the dig site based on Jace's interpretations?"

You can choose this question as a seed-question for your own investigation or come up with your own rabbit to chase!

Feel free to write down your question here:

Vague Notion: As you prepare to plan your investigation, first get a vague notion of what you plan to do—including guidelines on whether it will

involve experimentation, data-interpretation, modeling, arguing from evidence or some other practice of science.

The vague notion that we started from for our Learn by Example for this story is the following:

Go through a series of logic puzzles to figure out the principles of stratigraphy, including cross-cutting relationships and original superposition, and then use those principles to "reverse interpret" the geological setting of the dig site in Dinosaur Dig.

If you chose to go with our question for your own investigation, you can also choose our starting vague notion, or chase your own rabbit.

Feel free to write down your vague notion here:

Detailed Planning: Once you have your question and a vague notion in place, you can begin to fill in the nuances, complexities, and details of your plan. There is no recipe for your plan or a predetermined set of steps to follow, because your plan ***is*** the recipe or set of steps that you are creating for this particular investigation! Sketch out your investigation, anticipating as best you can what challenges you will encounter, and include as much detail as you are able (realizing that your investigative plan may not survive the first results of actually doing it.) Be sure to think about what background information you might need and how to get it, techniques you may need to research, the overall sequence of things you will need to do, and all other aspects of your investigation! You will likely need several pages to document your plan.

Follow a Question

Asking a question, choosing a vague notion of how to pursue that question, and then developing a detailed plan to pursue it is just the start of investigating. You actually have to do your investigation! During the course of doing your investigation, you will find out what works and what doesn't work. You may realize that you need more information, or different experimental tools, or even that your initial working model was so far off that your plan for investigation quickly falls apart. Authentic investigation involves iterative thinking during which your original experimental plan is revised, your initial interpretations updated, and your working model revisited. All of this is the whole point and practice of authentic investigation!

Keep careful notes on your experimental methods, interpretations, and models as well as your revisions as you work through your investigation.

Learn by Example (for Dinosaur Dig)

Our Starting Question: "Can we figure out the stratigraphy of the dig site based on Jace's interpretations?"

Our Starting Notion: Go through a series of logic puzzles to figure out the principles of stratigraphy, including cross-cutting relationships and original superposition, and then use those principles to "reverse interpret" the geological setting of the dig site in Dinosaur Dig.

Introduction:

A geological mystery lies at the heart of *Dinosaur Dig*. Jace's stepdad is digging up bones from the Cretaceous.

"So, we've clearly got dinosaurs from the early Cretaceous here, a lot of them. You can see this excellent specimen of a Sauropelta here."

However, they find a Stegosaurus in a layer above the Cretaceous material.

"But we dug up that piece of Stegosaurus last year above this layer. That doesn't make sense."

The Stegosaurus lived during the Jurassic time period, a time that came *before* the Cretaceous and so should *lie stratigraphically below it*!

What does this mean?

Understanding what they have found, and Jace's ultimate interpretation of it, requires us to understand stratigraphy! Let's try to figure it out!

Remember, the following puzzles and challenges are not intended to be systematic curricular steps through which students of science work toward defined learning outcomes. Rather, they are intended as aspirational examples of investigative thinking, appropriate at a variety of levels—some middle school, some high school, some for the expert teacher—to inspire investigative thoughts and increase understanding of what investigative thinking is all about. Find one or more that is right for you!

Challenge 1: Figuring out the Principles of Stratigraphy

Long before geologists could measure the age of a rock by radiometric dating, geologists figured out the stories in rocks (that is, the sequence of events recorded in the rocks) by determining the relative ages of rocks and features through consideration of key principles such as the following:

1) The principle of original superposition—the oldest sedimentary layer was originally on the bottom
2) The principle of original horizontality—sedimentary layers were originally deposited horizontally
3) Various cross-cutting relationships—preexisting things can be cross-cut by later events.

These are properly thought of as *principles* rather than *laws* because, rather than being the conclusion of scientific investigations in the way that laws are, they are logical reasoning steps that guide us in interpreting observational evidence.

What do we mean by logical? Well, it makes sense that a sediment deposited first will be (originally) on the bottom because gravity will pull it down and later sediments will then be deposited on top of it. It makes sense that a crosscutting event (that is, an event that affects a rock) had to happen after the affected rock already existed. I (Russ) often ask my students, "Here's broken pencil; which had to happen first, someone broke the pencil or someone made the pencil?" Clearly, the pencil has to exist before it can be broken. Rocks have to exist before they can be affected by later events such as erosion, magma intrusion, or faulting.

These principles seem very obvious and logical on the surface, but application of them is often more nuanced and complex than one might expect! These principles are at the very heart of geological investigation, and comprise a set of investigative methods that are somewhat unique to the geosciences. They provide a great opportunity to help students practice scientific reasoning and connect observations with reasonable interpretations.

Here are a few simple cross-cutting relationships to get started thinking about how they tell us about sequences of events in rock. The pictures below represent cross-sectional views of rock, that is, what we might see in the rock if we had a cut-away view down into the earth.

PUZZLE 1: In the cross-sectional view seen below, which came first, the rocks or the fault that cuts across them?

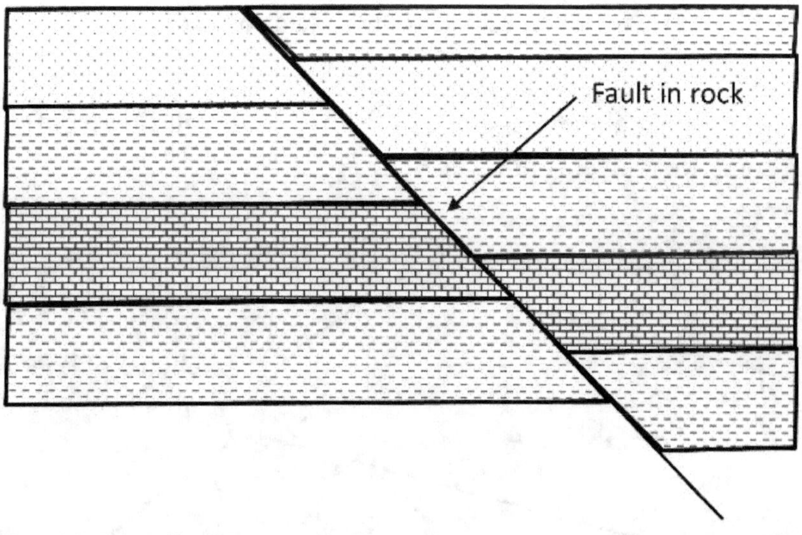

Spoiler Alert: Like the pencil, the rocks have to be there before they can be broken!

Puzzle 2: In the cross-section below, which had to come first, the igneous dike or the other rocks?

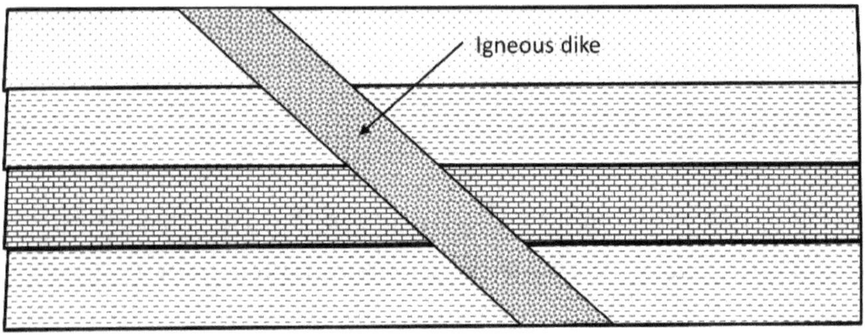

Spoiler Alert: We can see that the igneous dike cuts across the other layers, therefore the other layers had to be there first (it doesn't make sense that a big sheet of molten rock (later to become a dike) poked up into the air before the other rocks around it had formed).

Puzzle 3: What about the situation below? Can you figure out which had to come first, layer A or layer B?

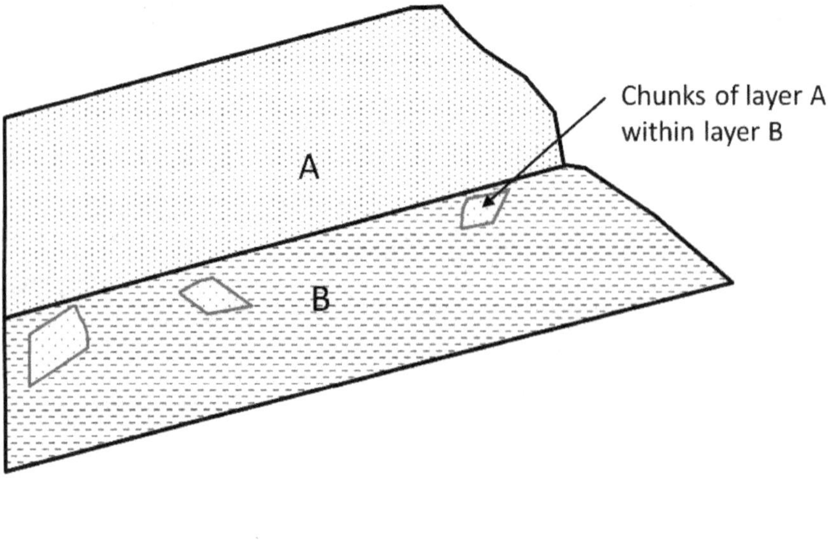

Spoiler Alert: We can see that chunks of layer A are found inside layer B. That means that layer A had to already exist when layer B formed, otherwise, we couldn't get layer A inside layer B (We can't put a pencil in our pencil bag until after the pencil already exists). Students might first think that layer B must be older because it is on the bottom, but yes, sometimes layers can get tipped upside down, like these two (although not commonly)! The inclusion of one rock inside another is not unlike the scenario in the story in which pre-existing fossils of Stegosaurus get included into a later rock layer (although according to the story, the rocks in the story are not tipped upside down).

PUZZLE 4: Consider the two different scenarios below. In the first, which is older, A or B? How do you know (that is, what is the cross-cutting relationships)? Can you tell if these rocks have been tipped upside down or not? How did B have to form?

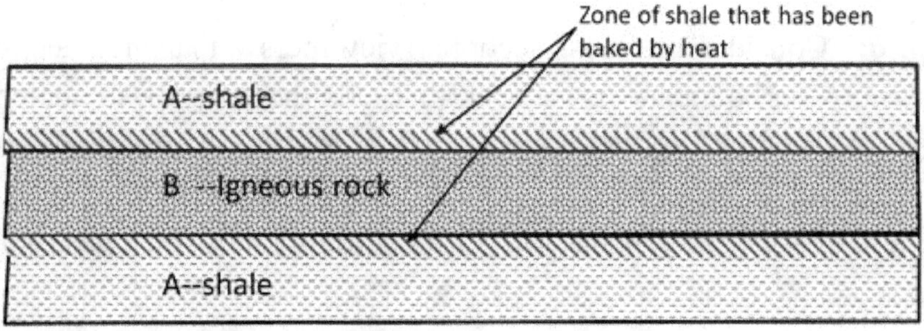

Spoiler Alert: The rocks on both sides of B had to already be there when B formed, otherwise they would not have been affected by the high temperature of the magma. This tells us that the magma intruded in between the pre-existing rocks. This feature is called an igneous sill.

PUZZLE 5: What about the case below? What is the age-order of A, B, and C? How can you tell? What must be different about how layer B formed in this case versus the case above?

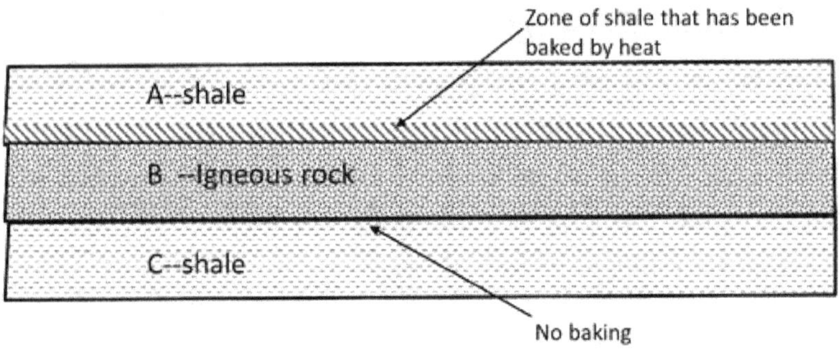

Spoiler Alert: In this case, only rock A was there at the time that B formed, telling us that B must have been a lava flow and layer C formed after the lava cooled. We can infer that these rocks have been flipped upside down relative to their orientation at deposition!

PUZZLE 6: Consider the cross-sectional view below that includes an unconformity—a boundary in the rock that tells the story of an ancient erosional surface that later got covered by new rock. Unconformities are often illustrated in cross-sections with a waving line, symbolizing the unevenness of the erosional surface. Can you figure out which has to be older, A or B?

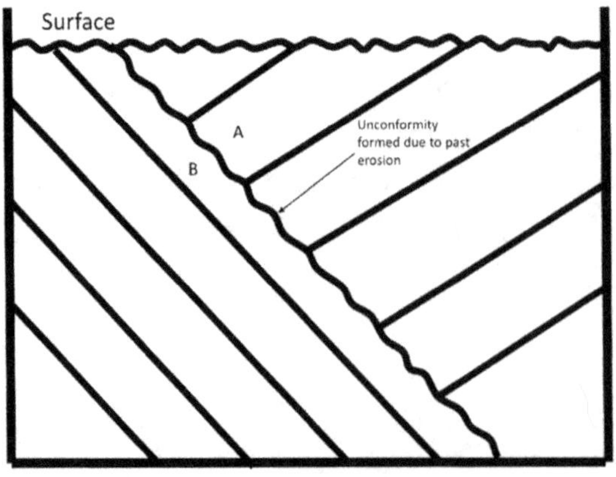

Spoiler Alert: Because the erosion surface cuts across layer A, but not layer B, we can infer that layer A had to be present before the erosion (you can't erode a rock layer that doesn't exist yet). Layer B was then deposited (horizontally, originally) on top of the old erosional land surface. Thus, A is older than B (once again, the rocks are tipped upside down—rare in real life, but common in our puzzles!)

Challenge 2: Applying Stratigraphic Principles to *Dinosaur Dig*

Now, taking the information from the story that we've focused on, let's figure out what the stratigraphy of the dig site must be like.

Remember:

"So, we've clearly got dinosaurs from the early Cretaceous here, a lot of them. You can see this excellent specimen of a Sauropelta here."

"But we dug up that piece of [Jurassic] Stegosaurus last year above this layer. That doesn't make sense."

PUZZLE 7: Which of the cross-sections below best represents the setting of the dinosaur dig in the story, where both Stegosaurus and Sauropelta fossils were found in a single layer of rock? Think about what you know from the story, which rock layers must be older, and which rock layers the fossils come from.

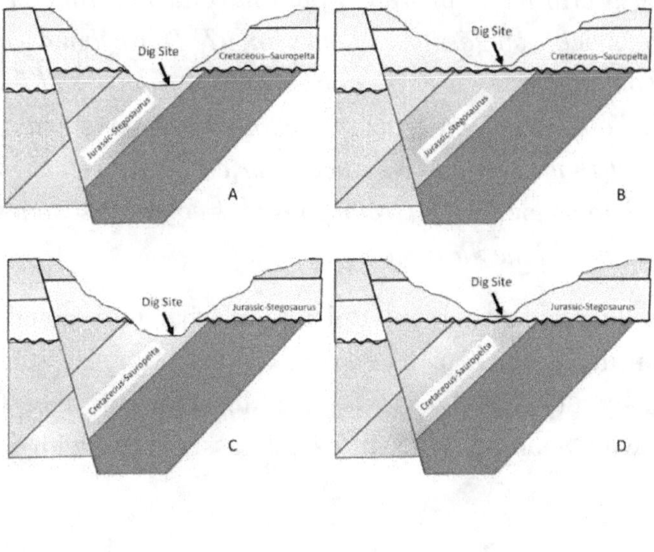

Spoiler Alert: This stratigraphic puzzle can be solved by recognizing that the bones were all found in the later (Cretaceous) rock, so the dig must be located in that layer, and by realizing that the Jurassic rocks had to form before the Cretaceous ones, thus the dig site in the Cretaceous rocks must be above and after the unconformity represented by the wavy line. That leaves us with option B. Option A is in Jurassic rock, where the later Sauropelta would not have yet existed and thus can't be included in the rock. Options C and D both have the Cretaceous being crosscut by the erosional unconformity, thus older than the unconformity, but yet the Jurassic rocks are younger than the unconformity, which is impossible since we have defined the Cretaceous to be the period that comes after the Jurassic.

PUZZLE 8: The core mystery of *Dinosaur Dig* is the following: how did the Jurassic Stegosaurus bones get mixed in with—and even above—the dinosaur bones from the much later Cretaceous?

While out looking for landscapes to draw, Jace takes note of a cliff with chunks of rock falling down to the ground below it, and has a revelation.

What if there had been an unstable wall of rock like this back in the Cretaceous, a cliff made of what would have at that time been ancient Jurassic rock with stegosaurus bones in it? Those bones might have weathered out and fallen just like chunks of rock from this cliff were doing today. The bones from the Jurassic could have gotten mixed in with Sauropelta at that time and then been buried in sediment where they'd slowly turn into a new rock, a Cretaceous rock containing bones from both the Cretaceous and the Jurassic.

Take some time and sketch out a picture of what Jace thought might be going on in the Cretaceous to explain the mystery. Be sure that your illustration—your model—shows the location of the Cretaceous sediments with Sauropelta fossils, the Jurassic rocks with Stegosaurus fossils, and makes clear how the strange stratigraphy of the fossils might have occurred. You might consider what the topographic landscape looked like when Stegosaurus bones were washing out of cliffs and mixing with Sauropelta remains, before the unconformity between the Jurassic and Cretaceous had finished forming, before the deposition of most of the Cretaceous rocks or any of the later rocks, and before the modern erosion that has exposed the dig site. Since you are modeling change through time, you might want two or three pictures in your model to represent different times during Earth's past. When you've finished, take a look at the next page to see our (Russ and Mary's) interpretation of what Jace realized might have happened.

Spoiler Alert: This situation is like Puzzle 3 above where a chunk of rock A included inside rock B tells us that rock A had to already be there when rock B formed. The fossils of Jurassic age inside Cretaceous age rock tell us that the Jurassic rock that contained the fossils had to already exist at the time the Cretaceous rock formed.

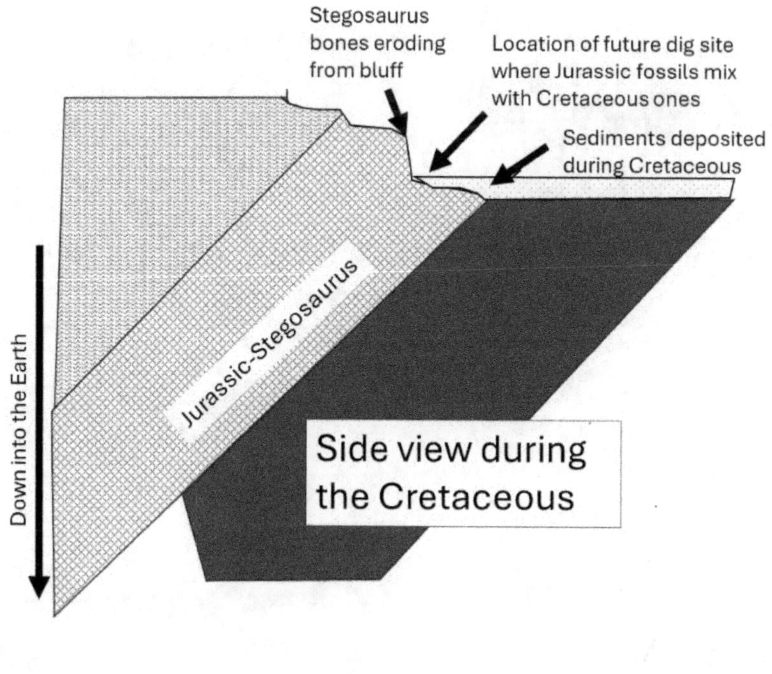

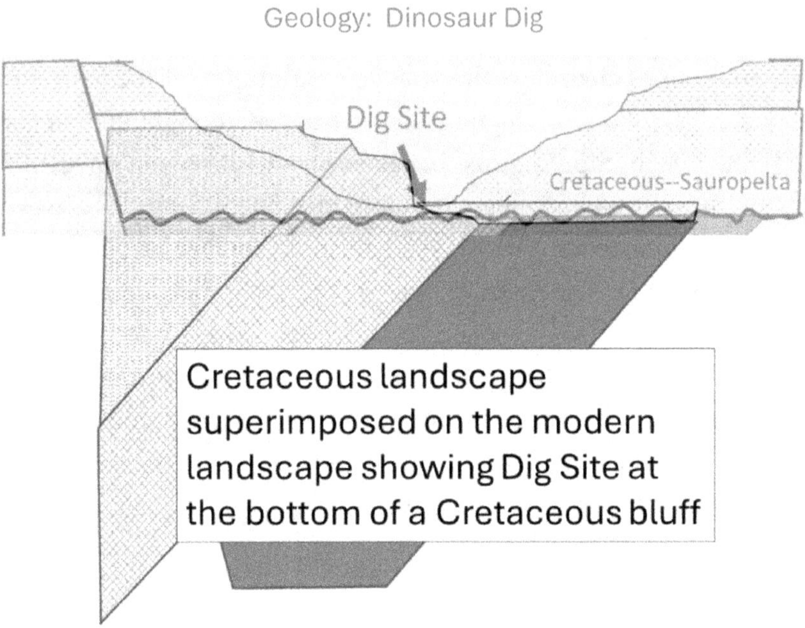
Geology: Dinosaur Dig
Dig Site
Cretaceous--Sauropelta
Cretaceous landscape superimposed on the modern landscape showing Dig Site at the bottom of a Cretaceous bluff

Reflection on Learning

The goal of this book is to practice investigation itself, not to focus on the information and ideas you understood better through investigation. If traditional learning outcomes are not the goal, then what have you learned? Taking some time to consider what you gained from your investigation, reflecting on the practice of science *metacognitively*, can help you hold onto what you have gained. For example: What questions did you ask? What observations and interpretations did you make? Did you *do* modeling, rather than just learning or applying a model?

Real investigation is fluid, creative, and defies easy categorization. Even so, it might be helpful to consider the following categories of investigation offered by the Next Generation Science Standard's science and engineering practices (NGSS Lead States. 2013. Next Generation Science Standards: National Academies Press).

- Asking questions and defining problems
- Developing and using models
- Planning and carrying out investigations
- Analyzing and interpreting data
- Using mathematics and computational thinking
- Constructing explanations and designing solutions
- Engaging in argument from evidence
- Obtaining, evaluating and communicating information

Notice that we have not numbered these practices. They are not steps in a sequential process, but rather components of an integrated and iterative process. Also recognize that this list of practices is **a** way to organize and understand the investigative practice of science, not **the** way. As you go through this reflection, be flexible in thinking about what aspects of your own investigation might be practices of science, because these practices can show up in unexpected ways. Remember, don't think so much about

what you learned or understood, but rather think about what you initiated, did, practiced or experienced.

On a separate sheet of paper, jot down some things that have strengthened your ability to pursue investigative thinking. Give examples of how you engaged in particular, specific practices.

Exploring with you the experience that we, the editors, had in developing our Learn by Example section might provide some guidance as you think about your own practices of science. Let's again start with practices and ask what aspects of our investigation used them.

As before, we certainly engaged with some traditional learning of the information and theories of science. In particular, the principles of stratigraphy—principle of original superposition, principle of original horizontality, and cross-cutting relationships—are a focus in this Learn by Example. However, learning information or theories was not our goal. Our goal was to practice investigation. The practice of science that we highlight in the Table of Contents for this Learn by Example is *interpretation.* In what way did we, the investigators who created and implemented this investigation, gain experience with this practice? Here are a few examples.

Puzzles 1-7 require **interpretation** of observational features in rock. The rocks themselves are the data in this case, and their characteristics reveal, through the use of the principles of stratigraphy, how they formed. For us to develop these puzzles, we had to do a kind of "reverse" interpretation in which we asked the question "What type of feature would result if a certain sequence of events transpired?" This process of reverse interpretation also required modeling of events and features as they developed through time.

Puzzle 8 requires readers to engage in a type of "reverse" **interpretation**, interpreting the implications of the structures seen in the modern-day rocks, but also modeling a sequence of events through time and envisioning how the rock layers would appear at different points in the past.

We also engaged with modeling, asking questions, constructing explanations, arguing from evidence, and other aspects of authentic investigation. To be authentic, your investigation need not engage in all the practices. However, real investigations often include many of these practices because the practices are not separate components but are parts of an integrated whole that we call scientific investigation.

Shifting Fortunes

By S. Q. Eries

Aki wants so much to stay in the home she's known her whole life, but with the Earth seeming to conspire against her, will she be able?

I gnaw my lower lip as Mama tallies our expenses. Her eyebrows bunched with concentration, she scribbles numbers, taps the calculator, and scribbles some more. Then her pencil falls, and relief floods her face. "Aki, there's enough left in the bank to last us another six months."

A whoop bursts from my lips. "Yes! We can stay!"

"For now. Whether we stay for good depends on the mushrooms." Her forehead puckers again. "I wish I didn't have to worry you with this. A fourteen-year-old girl should be having fun, not helping make ends meet."

I toss my head. "Who says I'm not having fun? You always called the mushroom shed my playground."

Mama's troubled look gives way to a small smile. "You really love it here, don't you?"

"It's home. And it'll stay that way." I take her hand in mine. "Don't worry. We'll be an overnight success at the farmers market. Soon all Japan will be raving about Kageyama Family Mushrooms!"

Mama chuckles, but worry still lurks in her eyes. We've gone through too much for her to get overly optimistic.

She glances at the kitchen clock. "You'd better go. You don't want to be late for school."

"Right." I grab my backpack. "And don't worry about the crates. I'll pack them as soon as I get home."

I race out of the house and through Mama's garden. At the end of the vegetable rows squats Papa's mushroom shed. Beneath its curved white roof are trays of jumbo champignon mushrooms, and stacked by the door are the brand-new crates we'll use to take them to market tomorrow.

As I wheel my bicycle from the side of the shed, I scan the forested slope that rises steeply behind the garden. Upon that hillside are slashed logs seeded with more of our mushrooms. Logs that Papa and I picked together.

He never imagined us turning his hobby into a livelihood. But he probably never imagined dying in a car crash either.

I swallow the lump in my throat and pedal off.

My bike zips through our woods to the narrow road that leads out of town. It's a winding ten kilometer ride to the next village where I catch the bus to school. As I shift gears for the steep climb, I pass the sign: Ohyu Geyser Welcomes You.

Tall weeds grow in an untidy tangle in front of the fading words. Our village used to maintain the sign with care, but no one bothers now.

After all, the geyser stopped flowing years ago.

I have only hazy memories of its three-meter high eruptions, but Ohyu Geyser was once the pride of the village. It attracted sightseers from all over, so many that the hot springs inn my great-grandfather built was always full. Then a decade ago, the geyser began to peter out. Its hourly eruptions became less frequent, less forceful.

The village tried everything it could think of to revive it. People made sure tourists didn't throw coins or trash into the geyser vent. Ohyu banned the drilling of new wells in the valley. Our inn even stopped pumping hot spring water for a month.

Nothing helped. The geyser's flow continued to decline. When it stopped altogether, so did the flow of tourists. Papa and my uncles struggled to keep Kageyama Inn open but, after five difficult years, closed its doors. Ohyu's few shops and restaurants followed suit. Soon most villagers, including my uncles, left for better opportunities. We stayed. Our family loved the mountain valley too much, and Papa found a lumberjack job that paid the bills.

But he's gone now.

I pedal furiously. Papa was determined to stay in Ohyu. That same determination burns in me. But despite scouring nearby villages, Mama can't find more than seasonal work. If the mushroom business fails–

I cut off that train of thought. *It has to work. It has to.*

***

A storm is drenching Hinata Village when the bus drops me off after school. As I duck beneath the convenience store awning to put on my poncho, a sleek sightseeing bus enters town. It halts before the manicured courtyard of Hinata Hot Springs Inn, and envy prickles my chest.

Tourists spill out. Elderly seeking relief from aches and pains. Hikers eager to ease sore muscles. City folk hoping to soak away stress. They scurry into the foyer, which already bustles with guests wearing Hinata Inn's indigo robes. As they enter, they pass a sign that advertises:

Hydrocarbonate Mineral Water

Indoor and Outdoor Pools

Years ago, our inn had a similar sign, and Kageyama and Hinata Inns were matched rivals for hot springs tourists. While Hinata was conveniently located near the expressway and rail station, we boasted a view of the geyser. But after its eruptions ceased, Ohyu became just another backwoods village, hardly worth the effort of visiting.

Why? I asked Papa that when our inn closed, and I remember the sadness that filled his eyes.

"Many villages have hot springs, but geysers are rare," he answered. "It takes a special formation and the right balance of groundwater and heat. Unfortunately, that balance is delicate.

Sometimes nature upsets it; sometimes people do. Either way, once a geyser stops, it's no simple matter to bring back."

A truck bearing the logo "Tanaka Water and Geothermal Wells" splashes by. I stare at the drill rig mounted in its bed and wish it could somehow reopen the geyser. But a geyser's water supply isn't at all like a pipe; it's a complicated network of underground cracks and fissures. And all it takes is a natural shift of the earth or mineral deposit buildup to disrupt that flow.

A hydrogeologist who visited Ohyu believed that was why our geyser stopped. But even if he was right, we on the surface can't possibly pinpoint those blocked pathways, let alone reopen them.

I unlock my bike and head home. Rain splatters my face, soaks my socks, and beats the pavement.

Of those billions of raindrops, some will evaporate when the clouds part. Others will join rivers, ultimately reaching the sea. Plants will take their share, drawing moisture through their roots. The rest will trickle into the earth. Some of that water will stay trapped in shallow soil; some will join the groundwater reserves that feed wells; yet another portion will continue down, descending to the depths where the scalding waters of the hot springs are born.

And once, a tiny fraction of that geothermal stream wound beneath Ohyu and emerged a spectacular jet, shooting to the sky like fireworks.

If only it would return. But that would take a miracle of geologic proportions.

***

It's still raining when I coast into Ohyu. As Kageyama Inn comes into view, I slow down. My uncles are trying to sell the property, and until they do, Mama and I keep an eye on it. The three-story wooden building is dark and silent as usual, but in the plaza beyond, I glimpse the bright colors of an umbrella.

My bike skids to a halt. "Granny Ito?"

The bent form beneath the umbrella waves, and anxiety buzzes in my chest. Of our few remaining neighbors, Granny Ito is the oldest. She always complains about arthritis and never goes out when it's wet. Hoping that her ninety-year-old wits haven't unraveled, I lean my bicycle against the inn's padlocked gates and trot to her.

She's sitting on a bench near the geyser's meter-high cone. Years ago, it spewed steaming spray to the delight of sightseers. Now only cold rainwater pools behind the safety barrier surrounding the silent vent.

Granny smiles. "Hello, Aki."

To my relief, her eyes are clear. "What are you doing, Granny? You'll catch cold."

"Don't worry. I'll go in soon. I just wanted to say goodbye to the geyser spirit."

A frog wedges in my throat. "Goodbye?"

"My daughter called today. She wants me to live in Osaka with her." Granny sighs. "I agreed."

"B-but you said you'd never leave the village."

"Things get difficult when you're old." Her gaze drifts to the inn's boarded-up windows. "This village never was large, but it was lively. At least it was when we had the geyser. With so few left now, my daughter worries."

I can't argue with that. Granny's the last resident on the main avenue. If something happened to her, the rest of us likely wouldn't realize until it was too late. As my heart sinks, she pats the spot beside her. "Sit with me."

We huddle beneath her umbrella. Before us, the geyser cone glistens in the rain. I think it looks eerie; chunky mineral deposits coat the cone and surrounding stones like an attack of mutant mold. Granny gazes like it's a long lost friend.

"I suppose it's silly to say goodbye. After all, the spirit left long ago." Her body sags. "We took it and its blessings for granted. Now we're paying the price."

She scans Ohyu's empty streets. "I wonder sometimes where it's gone and whether it might return."

I smile but say nothing. I know from school that geysers are steam explosions when superheated groundwater blasts to the surface. Spirits have nothing to do with it. But explaining that to Granny won't do a thing to bring the geyser back.

She pats my knee. "Enough of my rambling. Let's go."

I walk Granny to her house. The moment her door shuts, the smile on my face slips. I never imagined Granny would leave. Losing her is a huge blow.

I give myself a mental shake and mount my bicycle. I don't have time to be depressed. I have mushrooms to pack.

The next instant, the ground bucks like a horse, and I fall with a crash.

***

It was a 4.5 magnitude quake. Duration five seconds. Epicenter Ohyu Village. No casualties, no injuries. For most, it was barely newsworthy.

For us, it was a disaster.

I gape at the massive heap of dirt looming where the mushroom shed once stood. There's no sign of its white roof, not one mushroom tray, not a single champignon. Even the forest logs we seeded have disappeared beneath the landslide's path of destruction.

Mama weeps, her knees sinking into mud, but I'm too stunned to cry. The wooded slope behind our garden had always seemed so strong and enduring. Yet one shrug of the earth had brought it crashing upon the shed that sheltered our hopes.

Mama lurches to her feet. "That's it. We're finished."

I jolt from my daze. "Finished? But we can start over–"

She whips an arm toward the mound. "Everything's gone. It's a sign. We can't stay."

"But…"

She's not listening. Stony-faced, she marches into the house.

I crumple to the ground. Then the sobs come, and my tears mingle with the rain.

***

Two weeks later, the skies are clear above the village cemetery. Mama arranges fresh flowers while I clean the Kageyama Family headstone.

As I scrub off moss with an old wooden brush, guilt fills my chest. "I don't want to leave Papa."

"We'll visit," says Mama, not looking up. "He'll be fine. So will you. You'll love living with your cousins, and Tokyo will feel like home before you know it."

I doubt that. The few times we visited, I found the city stifling, and people sneered at our country accent. "Do we have to leave tomorrow?"

"We discussed this, Aki. I need to find work."

"Maybe you can go, and I'll stay a little longer."

Mama slams down the flowers. "With whom? Granny Ito's already moved, and I won't have you burdening the others. And if you say another word, we'll leave for Tokyo this instant."

Her words strike like an executioner's blade. These last days, even as we prepared to leave, I believed I could find a way to stay. Those last shreds of hope disintegrate.

Tears fill my eyes, and I bolt from the cemetery.

I race blindly down an overgrown path. Branches scratch my skin, but I scarcely notice. I'm desperate to get away from the inevitable, from my helplessness, from the hurt swallowing me alive.

My foot catches, and I go sprawling. Pain rips across my forearms and knees, and I lose all control.

"Give it back!" I scream. "Haven't I lost enough? Give it back to me!"

But there's no response, not even a sigh of wind.

I lift my head. I'm at the plaza. Through the haze of tears, the geyser cone looms, a gravestone marking the death of everything I've known, and rage surges anew.

"It's all your fault!" Scrambling up, I hurl the scrub brush in my hand. It smacks the crusty stones and drops with a splash.

Suddenly, a white jet bursts from the cone. I shriek, stumbling back, as Mama rushes into the plaza. "Aki!" she cries. "Are you–"

Her voice fades, and we gape at the column of vapor and water before us.

Seconds later, it disappears. I sprint across the plaza to the safety barrier. A pool of milky water steams within. From the size of it, the eruption wasn't today's first. From the thin vapor plume streaming out the vent, it won't be the last.

As my brain reels, Mama stammers, "Aki, was that real? How on earth?"

"The earthquake!" Temblors had the power to impact things above and below. The same jolt that triggered a landslide could also reshape the layers of soil and stone underfoot. And though that geologic shift crushed one dream, it may have opened the path to another.

Our gazes lock. For the first time in weeks, determination sparks in Mama's eyes. "Come on, Aki. We have plans to change."

***

S. Q. Eries is a registered professional engineer with master's degrees in civil engineering and environmental engineering. Her middle grade stories have appeared in *Cricket Magazine*, and she writes manga and book reviews for *The Fandom Post website*. She also writes YA fiction under the pen name Lynn Ng Quezon. Her YA novel *Mattie and the Machine*, a fictionalized account of inventor Margaret Knight's first patent, won the 2023 Grateful American Book Prize for excellence in literature about American history.

Art Credit: Shifting Fortunes artwork by Erin Colson.

"Shifting Fortunes" first appeared online in *Issues in Earth Science*, Issue 5, December 2015.

Chase Your Own Rabbits!

Question: Find something in the story that you wonder about, or want to challenge, or would like to explore more fully, and ask a scientifically-investigable question about it. As you consider things you wonder about, remember that not all questions can be investigated scientifically. For example, for the story *Shifting Fortunes,* a question related to how earthquakes might change the behavior of geysers could be interesting, but, to be investigative, it would need some component beyond getting answers from Google or an AI. A question like "what conditions will cause minerals in geyser water to crystallize out and make a cone" presents the problem that the SiO_2 content in geyser water, which makes up most cones, is so low that it takes thousands of years to form a significant cone—making it difficult to reproduce in the classroom, although you might use an analog material with higher solubility, like $CaCO_3$ which sometimes forms features in hot springs. Try to find a question that can be investigated scientifically with resources that you have, or can gather, in the classroom.

The question we, Russ and Mary, address in our Learn by Example for the story *Shifting Fortunes* is the following:

"Does changing pressure on water change its boiling temperature and can we relate that to geyser eruptions through modeling?"

You can choose this question as a seed-question for your own investigation or come up with your own rabbit to chase!

Feel free to write down your question here:

Vague Notion: As you prepare to plan your investigation, first get a vague notion of what you plan to do—including guidelines on whether it will

involve experimentation, data-interpretation, modeling, arguing from evidence or some other practice of science.

The vague notion that we started from for our Learn by Example for this story is the following:

The impact of the earthquake on the geyser is a key element of the story. We can work through some experiments with boiling water, apply that to an understanding of how geysers work (a model), and through that gain a better understanding of how the earthquake in the story could affect the geyser's behavior.

If you chose to go with our question for your own investigation, you can also choose our starting vague notion, or chase your own rabbit.

Feel free to write down your vague notion here:

Detailed Planning: Once you have your question and a vague notion in place, you can begin to fill in the nuances, complexities, and details of your plan. There is no recipe for your plan or a predetermined set of steps to follow, because your plan ***is*** the recipe or set of steps that you are creating for this particular investigation! Sketch out your investigation, anticipating as best you can what challenges you will encounter, and include as much detail as you are able (realizing that your investigative plan may not survive the first results of actually doing it.) Be sure to think about what background information you might need and how to get it, techniques you may need to research, the overall sequence of things you will need to do, and all other aspects of your investigation! You will likely need several pages to document your plan.

Follow a Question

Asking a question, choosing a vague notion of how to pursue that question, and then developing a detailed plan to pursue it is just the start of investigating. You actually have to do your investigation! During the course of doing your investigation, you will find out what works and what doesn't work. You may realize that you need more information, or different experimental tools, or even that your initial working model was so far off that your plan for investigation quickly falls apart. Authentic investigation involves iterative thinking during which your original experimental plan is revised, your initial interpretations updated, and your working model revisited. All of this is the whole point and practice of authentic investigation!

Keep careful notes on your experimental methods, interpretations, and models as well as your revisions as you work through your investigation.

Learn by Example (for Shifting Fortunes)

Our Starting Question: "Does changing pressure on water change its boiling temperature and can we relate that to geyser eruptions through modeling?"

Our Starting Notion: The impact of the earthquake on the geyser is a key element of the story. We plan to work through some experiments with boiling water, apply that to an understanding of how geysers work (a model), and through that gain a better understanding of how the earthquake in the story could affect the geyser's behavior.

Introduction

The Ohyu geyser had died, taking Aki's dreams with it. Then, one day, the geyser came back! What might have caused it to return?

"The earthquake!" Temblors had the power to impact things above and below. The same jolt that triggered a landslide could also reshape the layers of soil and stone underfoot. And though that geologic shift crushed one dream, it may have opened the path to another.

Is the on-again-off-again geyser that provides a key story element to *Shifting Fortunes* geologically reasonable? If so, why?

How does a geyser work, anyway?

The key to understanding a geyser is to understand the boiling temperature of water. Let's check it out!

Remember, the following puzzles and challenges are not intended to be systematic curricular steps through which students of science work toward defined learning outcomes. Rather, they are intended as aspirational examples of investigative thinking, appropriate at a variety of levels—some middle school, some high school, some for the expert teacher—to inspire investigative thoughts and increase understanding of what investigative thinking is all about. Find one or more that is right for you!

Challenge 1: Experimental measurement of the effect of pressure on boiling temperature

Deep underground in the plumbing system of a geyser, the pressure is higher due to the weight of all of the overlying water. Will the boiling temperature at that depth be greater than, less than, or equal to 100°C?

We can figure it out! Let's consider a graph showing possible ways that the boiling temperature depends on pressure. Take note that high pressure is lower on the y-axis to help us visualize that pressure increases with depth. Reversing the axis scale in this way might be initially confusing. However, this reversal forces us to recognize the graph's relationship to a real-world phenomenon, not just abstract math. The x-axis shows the temperature, and the various lines show possible ways that boiling temperature might depend on pressure.

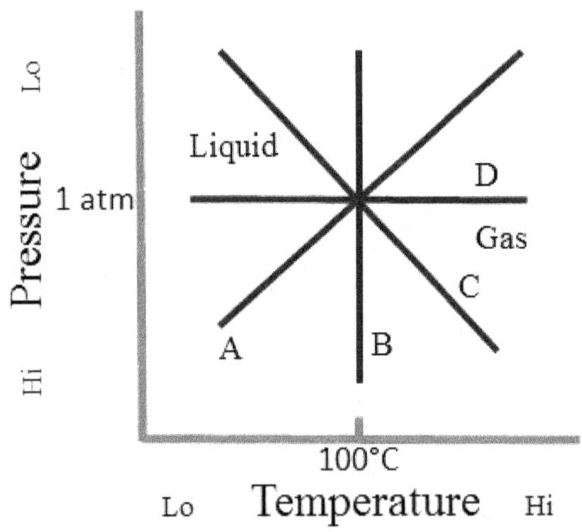

Which line on the graph best describes the behavior of water in nature?

We can determine which line is correct by making two measurements of boiling at two different pressures. To do these experiments, we need a stove, a pot, some water, a thermometer, a large syringe with no needle, and a balloon.

Measure the temperature at which water boils at 1 atmosphere pressure, using a thermometer in a pot of boiling water on the stove. If we are not at sea level, we might be able to measure that the boiling temperature is not exactly 100°C! Which way is it off? Is the pressure higher or lower than one atmosphere at that location?

A second measurement can be made by determining whether increasing or decreasing pressure will cause water to boil at room temperature. Put 20ml water in a 60ml syringe. Make sure all air is removed from the syringe. Hold a balloon over the end of the syringe with your thumb to seal it up. Then alternately pull or push on the syringe (decrease or increase pressure in the syringe), and see which works to boil the water (you can see vapor bubbles start to form inside the water when boiling starts). Plot our two points on the graph. From this, we can determine which graph line must be correct.

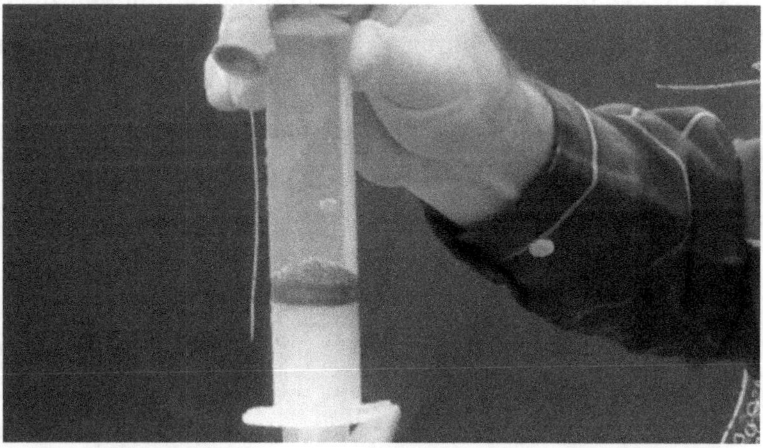

Water boiling at room temperature when pulling down on the syringe plunger.

So which line in the graph above is correct?

a. Boiling temperature decreases as pressure increases, like line A
b. Boiling temperature stays constant as pressure changes, like line B
c. Boiling temperature increases as pressure increases, like line C
d. Boiling temperature changes even if pressure stays constant, like line D

Spoiler Alert: If the boiling temperature changed erratically even as pressure stays constant, like D, that would be a strange and chaotic world indeed! The reason that science works is that the universe is not chaotic, but is regular in its behavior, and therefore predictable. We know from common experience that boiling temperature changes with pressure, explaining, for example, why recipes call for different amounts of cooking at different elevations, thus eliminating line B. From the experiment with a syringe (or looking at the picture of Russ doing it) we see that there exists some pressure less than one atmosphere where water will boil at room temperature. Thus, decreasing pressure decreases boiling temperature, or increasing pressure increases boiling temperature, line C.

Challenge 2: Building a conceptual model

A second key to understanding geysers is to think about what is going on below the ground to cause geysers and then what happens to cause a geyser to stop its eruptions. Unlike hot springs, geysers have a constriction in their piping which prevents circulation of heat from deeper in the geyser to the surface. Since heat can't escape from the deeper parts of the geyser system, water deeper in the geyser system slowly gets hot enough to boil. These plumbing systems are shown below, comparing a geyser with a hot spring.

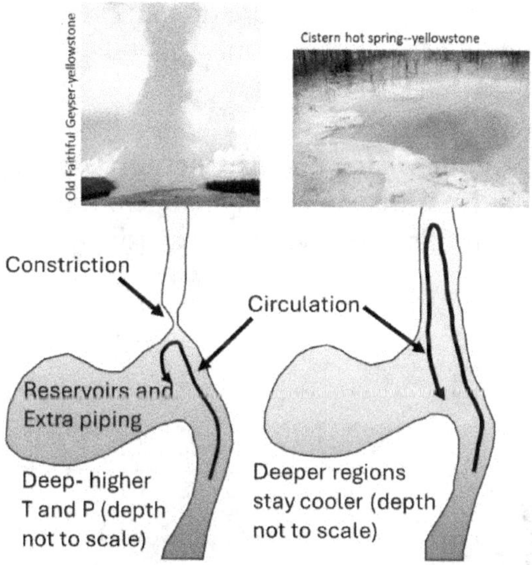

Given this conceptual model for geyser plumbing shown above, consider the steps below that describe one conceptual model for how a geyser works.

1) Constrictions in the piping allow heat from within the Earth (often originating from nearby magma or very hot rock) to slowly raise the water temperature in the lower part of the geyser plumbing until it reaches boiling temperature.
2) When the water starts to boil, the rising and expanding water vapor expels water out at the surface, which decreases the pressure lower down, which decreases the boiling temperature lower down, which causes more boiling and more throwing water out at the surface (the geyser erupts). Coincident with this process, heat is lost to the water vapor escaping to the surface, thus decreasing the water temperature as pressure falls but keeping temperature at the boiling point for the pressure.
3) Once the water cools off enough (or we run out of water), the boiling stops, and the geyser plumbing slowly fills back up with cool water, causing pressure to rise again.

Let's match up each of the three steps above (and abbreviated below) with one of the arrows on the temperature-pressure diagram (below—this diagram shows the boiling temperature of water). Be sure to think about what the temperature-pressure graph means!

1) Water temperature rises toward boiling point.
2) Water boils, and the rising and expanding water vapor expels water decreasing both pressure and boiling temperature.
3) Geyser plumbing fills with cool water, causing pressure to rise.

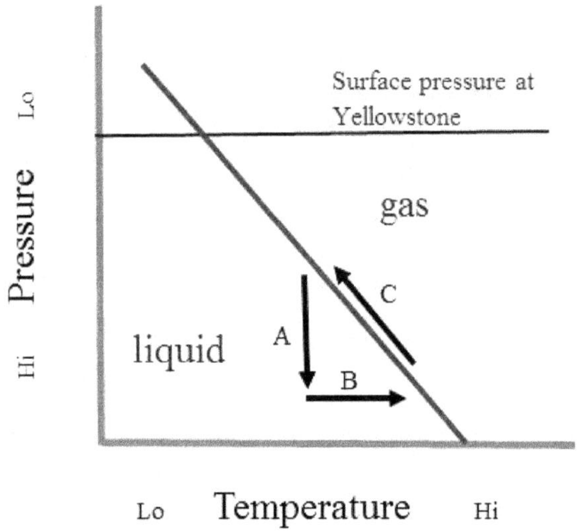

Spoiler Alert: 1) Heat gradually builds at depth causing temperature of the water there to rise. Because of the high pressure, the water doesn't boil yet. This is shown by arrow B. 2) The water starts to boil, and the rising water vapor throws water out of the top of the geyser, which decreases pressure, which causes even more boiling because water boils at a lower temperature when pressure decreases. The water vapor takes heat with it, causing the temperature to decrease as pressure decreases, staying on the "boiling curve". This is shown by arrow C. 3) Once we run out of either heat or water, the boiling stops, and the geyser reservoir has to fill back up with water, and the pressure gradually rises. This is shown by arrow A.

Challenge 3: What did the earthquake do to restart the geyser?

So, what might the earthquake have done in *Shifting Fortunes* to cause the geyser to start working again?

Spoiler Alert: Probably the earthquake changed the underground plumbing such as to allow the cycle of heat build-up and eruptive boiling to resume, perhaps by causing a constriction to re-form where it had been more open, or reopening a fully closed pipe to the surface.

Reflection on Learning

The goal of this book is to practice investigation itself, not to focus on the information and ideas you understood better through investigation. If traditional learning outcomes are not the goal, then what have you learned? Taking some time to consider what you gained from your investigation, reflecting on the practice of science *metacognitively*, can help you hold onto what you have gained. For example: What questions did you ask? What observations and interpretations did you make? Did you *do* modeling, rather than just learning or applying a model?

Real investigation is fluid, creative, and defies easy categorization. Even so, it might be helpful to consider the following categories of investigation offered by the Next Generation Science Standard's science and engineering practices (NGSS Lead States. 2013. Next Generation Science Standards: National Academies Press).

- Asking questions and defining problems
- Developing and using models
- Planning and carrying out investigations
- Analyzing and interpreting data
- Using mathematics and computational thinking
- Constructing explanations and designing solutions
- Engaging in argument from evidence
- Obtaining, evaluating and communicating information

Notice that we have not numbered these practices. They are not steps in a sequential process, but rather components of an integrated and iterative process. Also recognize that this list of practices is **a** way to organize and understand the investigative practice of science, not **the** way. As you go through this reflection, be flexible in thinking about what aspects of your own investigation might be practices of science, because these practices can show up in unexpected ways. Remember, don't think so much about

what you learned or understood, but rather think about what you initiated, did, practiced or experienced.

On a separate sheet of paper, jot down some things that have strengthened your ability to pursue investigative thinking. Give examples of how you engaged in particular, specific practices.

Exploring with you the experience that comes from this particular Learn by Example might provide some guidance as you think about your own practices of science. For the preceding stories, we discussed what aspects of science practice that we, the editors, experienced in developing the Learn by Example investigations. This time, let's consider what practices of science that a reader who works through the offered puzzles might experience. The investigative experience is more robust for the person who develops and implements an investigation compared to the person who simply works through an already-developed investigation. However, some elements of the practices of science can still be experienced.

The reader can find some traditional learning outcomes, for example, understanding concepts of how geysers work (geology of the underlying rock, chemistry of phase transitions, physics of heat flow). However, learning information or theories is not our goal. Our goal is to practice investigation. In what way does someone who works through the puzzles of this investigation gain experience with the practices of science? Here are a few examples.

You might think that the person who simply works through the puzzles of this investigation does not really engage with developing an experiment, and this is partly true. However, working through the puzzles engages the reader in **interpreting** the results of the experiment, that is, what the water boiling in the syringe at room temperature means, and **interpreting** the Challenge 1 graph in terms of real-world phenomena.

Also, the application of the experimental results to a graph requires **mathematical thinking**, figuring out what water boiling at 100°C at 1 atm pressure and water boiling at room temperature at some lower pressure means in terms of the lines on the graph, providing, for example, a visceral understanding positive and negative slopes.

You might think that because the model for how a geyser works is presented in the text, the person who simply follows this investigation is not involved in model development, and this is partly true. However, working through the challenges requires detailed consideration of how pressure, temperature, water volumes, and water movements are coordinated so as to result in a geyser eruption. This requires extensive mental **modeling** of the process. Also, creating an explanation for the final question—What did the earthquake do to restart the geyser?—requires active development of a quite different **model** related to how the earthquake might have changed the factors that our other model suggests affect the geyser.

Ideally, the person working through these puzzles attempts to answer the questions, and not simply read on to "find the answer." If so, they will engage in **explaining observations** and **arguing from evidence**. Hopefully, also, the curious reader **asks questions** of their own, perhaps **obtaining and evaluating information** to address those questions!

Notice that the only practice not present is "planning and carrying out investigations!" It is perhaps the hardest of the NGSS practices to experience by working through someone else's investigation. First-hand experience with this practice requires an active engagement in developing and doing the investigation. Nonetheless, remember that your investigation need not engage in all the practices to be authentic.

Story by Mary Ibe

A True Original

By Mary Ibe

Young urban explorer LaRose becomes a true original when curiosity and imagination involve her and Grandpa Deon in a real environmental science investigation.

Oh, wow! This is a really good one!" LaRose said, pointing to a wet mound. It looked like a tiny, lumpy smokestack.

"Yeah, that is a nice one, Rosie-bug," Grandpa Deon said. "You always have been good at finding chimneys."

LaRose smiled, feeling too old to still be called "Rosie-bug," but somehow not minding that it was her beloved grandpa calling her that.

LaRose and Grandpa Deon were keeping up their tradition of hunting for crayfish chimneys after the spring rains. Crayfish chimneys aren't like the chimneys that go with a fireplace. A crayfish chimney forms when a crayfish digs a tunnel, making little pellet-sized balls of mud out of the soil. They stack these mud pellets outside of their tunnels and the piles end up looking like little chimneys.

"Hey, did I ever tell you about my prize-winning crayfish chimney collection?" Grandpa Deon asked.

"What?!" LaRose said, with surprise. "You mean to tell me that we've been hunting for crayfish chimneys every spring and you are just now telling me this? How could you hold back that kind of information from your favorite Urban Explorer?!?" She playfully tapped his arm for emphasis.

Grandpa Deon and LaRose referred to themselves as Urban Explorers. Even though they lived on the edge of a large city, they still looked for nature wherever they could find it. The drainage ditch that ran along the sidewalk that stretched between their two apartment buildings supplied hours of fun for them.

"Come on over to my workshop and I'll show it to you," Grandpa Deon said.

The two of them made the short trek to what Grandpa Deon called his workshop. It was really the maintenance shed for the apartment complex. Grandpa Deon got a reduced rate on his rent in exchange for doing odd jobs and fixing things in the apartments. He unlocked the shed door and they walked in. Stacked up on a high shelf was an assortment of jars with lids on them. Grandpa gently took one of the jars down and dusted off the lid and the sides of it and handed it to LaRose.

LaRose peered into it and slowly turned the jar around. “Yeah, now I see it. It is definitely a crayfish chimney. How old is it, do you think?”

“It must be fifty years old. I collected about twenty chimneys for a school assignment when I was about your age. We had to collect something and learn how to classify it. Most kids collected leaves or feathers or stones. But, even back then I loved crayfish chimneys. I decided that I would dig some up and save them in jars. I let them air dry and I labeled them based on where I found them and arranged them according to size and location.”

“But you said they were prize-winning. What did you win a prize for?”

“Originality!” Grandpa Deon bellowed with mirth. “My teacher said nobody had ever submitted crayfish chimneys for the assignment. See, there’s the blue ribbon she gave me,” he said, pointing to a faded blue ribbon with its tail tucked under one of the jars.

LaRose gently took the ribbon from the shelf and read the hand-lettering on it. It said, “For Deon, A True Original!”

Seeing the look of pride Grandpa still had when he looked at the ribbon made LaRose smile.

The next day after school, LaRose again met Grandpa at the drainage ditch. “Hi Grandpa, I’m going to be an original. I’m going to collect crayfish chimneys, too!”

“Now how can you be original if I already did that?” he asked.

“Well, I’m not only going to collect them, I’m also going to analyze them,” LaRose said.

"What will you analyze?" Grandpa asked.

"I haven't figured that part out yet, but I'll think of something," she answered.

"Well, you should think of something quick, because it seems like either the crayfish chimney season is really short this year or there are way fewer crayfish than there used to be."

"Hey, that's it!" LaRose said. "There aren't as many crayfish as I remember from when I was really little. I can analyze them to figure out why that is."

"From what I remember, there were not only more crayfish, but the area where I would find chimneys as a boy was much larger back then. Now it seems like we mostly find them in this little part of the creek that bends away from the street."

LaRose looked up at the banks of the little creek that had provided so many hours of entertainment. It was covered in trees, roots and shrubs, all sprouting new spring leaves. Water dribbled from a pipe that drained the water off of the road. It made a pretty sound, like a little waterfall, as it fell into the drainage ditch. She sighed with contentment. "I just love this place," she said. "But now I'm worried. We are definitely onto something about the crayfish disappearing. I wonder what it could be?"

"I don't know what's going on, but we are Urban Explorers, so let's investigate!"

"Ok. Let's start by seeing how far the area where we can find crayfish chimneys goes," LaRose suggested.

"It seems like it stops right here," LaRose said, pointing to an area that ended close to the little drainpipe waterfall from the road. "Grandpa, when you were a kid, how far could you go and still find chimneys?"

"Well, that would be hard to compare because when I was a boy, that street didn't even exist," he said, pointing up the bank to the nearby road.

That night LaRose searched the internet for ideas on how to find out more about the disappearing crayfish chimneys. She didn't discover any answers, but she did come across a soil test kit that looked interesting. She could get it for free from a county agency. She filled out a form online and a few days later, she got a test kit in the mail. She followed the instructions and mixed samples of chimney soil with water and tested the mixture for sulfates using a strip of paper. All she had to do was see if the paper changed color and report the paper color and location of the soil sample to the agency.

LaRose opened the county agency website and began to answer the questions it asked her about her results. She hit submit on her responses and went to bed.

The next morning, she noticed a notification in her inbox. It was from the county agency. They said that her test results were very interesting, and they wanted to know if she could get any more test samples from the same area. LaRose gathered up as many old jars as she could find in her recycling bin and went to collect samples to test.

She submitted the results of her new test samples that same day. Just like before, the very next day, she had a message from the county agency, asking for more samples. She was out of jars, but she knew where she could get some more.

"Grandpa!" she called, swinging open the door to his workshop. "Is there any chance I could use some of the soil from your crayfish chimney collection? I only need a couple of spoonfuls from each."

She explained to him why she needed them. He agreed to let her scrape some soil from each of his jars. He also gave her his category sheet so she could report to the county agency the location of each chimney.

As LaRose looked at his location list, she said, "Hey, some of your chimneys are in almost the same exact location as the samples I already submitted. I wonder if the results will be the same?" She completed the testing and became worried when some of the paper strips from Grandpa Deon's samples turned a different color than the samples she gathered the first time. When she submitted Grandpa's results, she mentioned the color difference and asked the county agency if there were any further types of analyses they could do on all the samples to explain the difference.

Two days later she got another message from the county agency. It said: Would you be able to send us actual samples of soil for further analysis? We would like to investigate further. Also, we noticed that you submitted some sample locations twice. Could you please supply the dates you collected the samples?

Their request caused LaRose to realize that the county agency seemed to be very eager to learn more about her samples. What did they find so interesting about her samples? Had she stumbled across something important? She had to find out more. Maybe it would somehow help the crayfish. So, instead of just answering their questions, she decided to also ask them a few questions.

In her response she asked them, “Is there something unusual about the samples I submitted? Do you think there is something in the soil affecting the crayfish? I don't know the exact date for all of them, but my best estimate is that the most recent batch of results I sent to you was from fifty years ago. The first batch I sent you was May 22 and the second batch was from May 24.”

The very next day, the county agency sent her a short response with a question: “May 22 and 24 from fifty years ago?”

“No. From this year,” LaRose typed and hit send.

She immediately got a notification from them. It said, “Can we call you?”

“Give me five minutes,” she responded and gave them Grandpa Deon’s cell phone number.

LaRose ran to her Grandpa’s workshop and opened the door with a shout, “Grandpa, the phone’s going to ring, please answer it!”

Just then, his phone rang.

“Hello?”

“Hello, may I speak with LaRose Taylor please?” said a voice.

“May I ask who is calling?”

“This is Shari from the County Water Agency. LaRose has submitted some soil samples to us that we would like to speak to her about.”

Grandpa looked wide-eyed and passed the phone to LaRose.

"Hello," LaRose said, trying to sound confident.

"LaRose, what a pleasure to speak with you. I am Shari, the researcher in charge of the Citizen Science Soil Sample Project you have been submitting test results to. Did you know the samples you collected are telling quite a story?"

"What do you mean?" she asked excitedly.

"We think that the chemicals in the soil are impacting the whole ecosystem, not only the crayfish. What changes have you noticed?"

"We have noticed that there doesn't seem to be as many crayfish chimneys the last few years," LaRose said.

"Your test results indicate that now there are large amounts of chemicals in the soil that could be harmful to plants and animals. The samples from fifty years ago do not show evidence of any of these chemicals. Parts of the geochemical fingerprint of the soil have changed over time and we want to learn more."

"I don't know what that geochemical fingerprint thing is you mentioned, but I can tell you that there is a road near the creek that wasn't there fifty years ago," LaRose said.

"We know about that road and the data from your samples confirm our suspicions. We want to do some more tests to look at the soil's fingerprint. The fingerprint, or the types of chemicals that are in the soil in your samples, is intriguing. Your soil samples look like they might be a key piece of information for our project. Not many people have such an interest in crayfish chimneys. You are a true original, LaRose! Can we meet you sometime and look at your samples?" asked Shari.

LaRose talked to Grandpa Deon and the three of them worked out a time and a place to meet.

"Your curiosity is really something, LaRose," Grandpa Deon said. "Just imagine, your soil samples are going to help all the wildlife in the area!"

Grandpa Deon reached up and took his prize-winning blue ribbon off the shelf and said, "Here LaRose, you deserve this ribbon. Even the county agency thinks you are a true original!"

***

Mary Ibe is a writer who has been enamored with writing fiction ever since winning a Scholastic Writing Award in ninth grade. While a middle school science teacher for twenty years, Mary would get her writer's fix by having stories published here and there in periodicals such as *Girl's Life* and *Primary Treasure*. She currently writes middle school STEM curriculum for Project Lead The Way, an organization that provides project-based curriculum. After visiting an impressive Rock and Mineral Museum on vacation, Mary knew it was time to get back into writing fiction, which is how this story on geochemical fingerprinting emerged.

Art Credit: A True Original artwork by Autumn Williams.

"A True Original" first appeared online in *Issues in Earth Science*, Issue 20, December 2025.

Chase Your Own Rabbits!

Question: Find something in the story that you wonder about, or want to challenge, or would like to explore more fully, and ask a scientifically-investigable question about it. As you consider things you wonder about, remember that not all questions can be investigated scientifically with classroom resources. For example, for the story *A True Original,* a question like "how long can pollutants remain in the soil when rains might wash them out" might be explored experimentally, but you would need to choose a pollutant analog for investigation that would be safe for classroom use and would be sufficiently mobile in rainwater to allow you to measure the effect within the timeframe of a classroom investigation. You might wonder exactly what "chemicals in the soil" can be related to the nearby road AND would affect the crayfish population—a question that would be difficult to answer in the classroom with either field or experimental investigation, but you might make a careful comparison to analogous studies that you locate online. Try to find a question that can be investigated scientifically with resources that you have, or can gather, in the classroom.

The question we, Russ and Mary, address in our Learn by Example for the story *A True Original* is the following:

"Even if there are differences in the chemical analyses from one sample to another, how can we know that those differences are real or meaningful and not just chance variations?"

You can choose this question as a seed-question for your own investigation or come up with your own rabbit to chase!

Feel free to write down your question here:

Vague Notion: As you prepare to plan your investigation, first get a vague notion of what you plan to do—including guidelines on whether it will involve experimentation, data-interpretation, modeling, arguing from evidence or some other practice of science.

The vague notion that we started from for our Learn by Example for this story is the following:

Statistical thinking is an essential part of all investigation in science. Although we are not given statistical information about the samples in A True Original *(or any data at all), we can imagine various scenarios and place some constraints on what the data must have looked like for the county scientists to be intrigued.*

If you chose to go with our question for your own investigation, you can also choose our starting vague notion, or chase your own rabbit.

Feel free to write down your vague notion here:

Detailed Planning: Once you have your question and a vague notion in place, you can begin to fill in the nuances, complexities, and details of your plan. There is no recipe for your plan or a predetermined set of steps to follow, because your plan ***is*** the recipe or set of steps that you are creating for this particular investigation! Sketch out your investigation, anticipating as best you can what challenges you will encounter, and include as much detail as you are able (realizing that your investigative plan may not survive the first results of actually doing it.) Be sure to think about what background information you might need and how to get it, techniques you may need to research, the overall sequence of things you

will need to do, and all other aspects of your investigation! You will likely need several pages to document your plan.

Follow a Question

Asking a question, choosing a vague notion of how to pursue that question, and then developing a detailed plan to pursue it is just the start of investigating. You actually have to do your investigation! During the course of doing your investigation, you will find out what works and what doesn't work. You may realize that you need more information, or different experimental tools, or even that your initial working model was so far off that your plan for investigation quickly falls apart. Authentic investigation involves iterative thinking during which your original experimental plan is revised, your initial interpretations updated, and your working model revisited. All of this is the whole point and practice of authentic investigation!

Keep careful notes on your experimental methods, interpretations, and models as well as your revisions as you work through your investigation.

Learn by Example (for A True Original)

Our Starting Question: "Even if there are differences in the chemical analyses from one sample to another, how can we know that those differences are real or meaningful and not just chance variations?"

Our Starting Notion: Statistical thinking is an essential part of all investigation in science. Although we are not given statistical information about the samples in A True Original *(or any data at all), we can imagine various scenarios and place some constraints on what the data must have looked like for the county scientists to be intrigued.*

Introduction

LaRose, our young protagonist in *A True Original*, is a natural investigator. She starts out with an interest in the world around her, noticing the crayfish burrows and collecting them. She comes up with a question to address—why the number of crayfish chimneys has declined—and an investigative approach to answer the question—analyzing her chimneys and the older chimneys of her grandfather. When the county finds a difference between the old and new samples, she even begins to construct models to explain the change.

..."The samples from fifty years ago do not show evidence of any of these chemicals. Parts of the geochemical fingerprint of the soil have changed over time and we want to learn more."

"I don't know what that geochemical fingerprint thing is you mentioned, but I can tell you that there is a road near the creek that wasn't there fifty years ago," LaRose said.

But, how does the county know that the samples from the past are different from the more recent samples? How can they tell real differences related to the age of the samples from chance variation between different samples that might not be related to age at all?

Let's address the important idea of statistical uncertainty. In science, no measurement is ever exact, and so scientists need to account for how much uncertainty there is in their measurements. In addition to measurement (or analytical) uncertainty, uncertainty can arise from true variation in real-world samples (no two samples will ever be exactly alike, even if taken from the same area). How can scientists take this variation and uncertainty into account when comparing different samples, like the comparison between the modern and older crayfish chimney samples in the story *A True Original*?

Often, scientists use statistical analysis, deriving values such as standard deviation, to facilitate comparisons. The mathematics involved in this process often discourages teachers from introducing their students to this critically important idea of science, which leaves a giant hole in the students' understanding of the nature of scientific inquiry. However, an understanding of statistical uncertainty does not require a lot of math or equations! A colleague of Russ's who worked extensively with statistical variation within the context of real Moon samples and his favorite analysis method (Instrumental Neutron Activation Analysis, or INAA) once said, "If you can't see the difference on a graph, then all the statistical calculations in the world aren't going to convince people that a variation is real."

So, let's think about understanding statistics using graphs!

Remember, the following puzzles and challenges are not intended to be systematic curricular steps through which students of science work toward defined learning outcomes. Rather, they are intended as aspirational examples of investigative thinking, appropriate at a variety of levels—some middle school, some high school, some for the expert teacher—to inspire investigative thoughts and increase understanding of what investigative thinking is all about. Find one or more that is right for you!

Challenge 1: Initial puzzle and conceptual analysis

The story, *A True Original*, talks about measuring some type of "sulfates." Sulfur present as sulfides can affect pH, which might affect environment, and, in addition, many toxic metals, like arsenic, might be present in association with sulfur.

PUZZLE 1: Let's suppose that we analyze the arsenic sulfate concentration in a sample of the modern chimney and compare it to a sample of an older chimney. Suppose that the analyses of the "old" crayfish chimneys gave a result of 10 and a sample from a "new" chimney gave a result of 12. Are these different? What do you think?

Spoiler alert: The best answer is: "I have no idea from this information!" There are a lot of problems with the question, "are these different?" First of all, we are given no units. 10 or 12 what? Are we comparing apples to apples or apples to oranges? In science, numbers have little meaning until they are associated with a unit. For example, perhaps it is 10 *percent* arsenic sulfate and 12 *parts per million* arsenic sulfate. Clearly, the 10 is much, much bigger (by 4 orders of magnitude) than the 12!

PUZZLE 2: Let's suppose they are both in the same units, say parts per million. Now, can you say that one is different from the other?

Spoiler alert: Still no! From these data you have no understanding of either the real-world, natural variation of arsenic sulfate or the statistical uncertainty associated with the measurements. For example, perhaps the real-world natural variation is large enough that the 12ppm measurement is not meaningfully different from 10ppm measurement. Regarding the analytical uncertainty, maybe the best that you can do with measuring is to get within about 5 parts per million of the true value. So, one time you might measure 10, but the next time it might be 5 and the time after than 15. You can see that from the reported data of 10 and 12 alone, you can't know whether or not 10 is meaningfully different from 12.

PUZZLE 3: Suppose that I determine the *average* value of 5 analyses for each of the old and new samples. If the average concentrations come out at 10ppm and 12ppm, what can I say about whether they are different?

Spoiler alert: Still, nothing. Without knowing the values of the *individual analyses*, the average provides no additional useful information! We don't know anything about the range in values or how many values contributed to the average. When students take an average and then throw away the individual values, they have just thrown away their most important data!

Challenge 2: Plotting and interpretation of a single chemical component

So, what is one to do? One method of assessing uncertainty is to compare multiple analyses to each other and/or compare multiple samples to each other. Comparing multiple analyses of the same sample allows one to determine analytical uncertainty, and comparing multiple samples allows one to evaluate real variation between samples.

Consider the following sets of data that (in our pretend scenario) were derived from the crayfish chimneys. To evaluate the uncertainty in a graphical way, plot these data along a line (see the line for each scenario below), using different symbols for the old and new chimneys, and then say whether you think that the values: 1) are different, 2) might be different but we can't be sure, or 3) show no indication of being different.

PUZZLE 4: Scenario 1 with data for arsenic sulfate.
Old Chimneys: 8ppm, 9ppm, 6ppm, 20ppm, 10ppm, 13ppm, 18ppm
New Chimneys: 15ppm, 10ppm, 17ppm, 7ppm, 12ppm, 9ppm, 14ppm

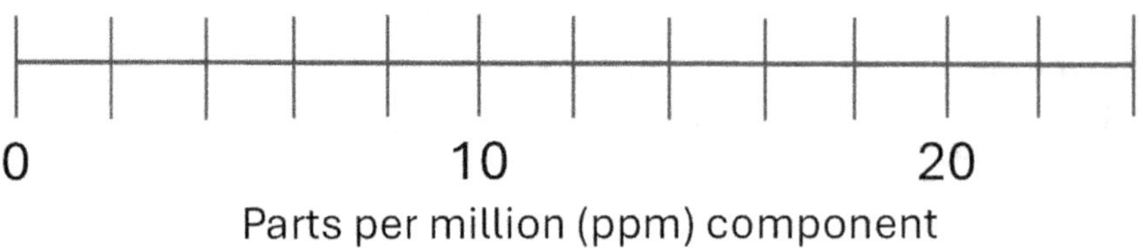

Interpretation goal: Based on our graph, are the values different or not? Frame your interpretation as one of the following: the values are different, the values might be different but we can't be sure, or the values show no indication of being different. The analysis should include an explanation of <u>why do we think that.</u>

Spoiler alert: For this problem, the average concentrations are exactly the same (12ppm), making the interpretation that they are not obviously different...rather obvious. However, the variation in values is quite large relative to the average concentration and so the 'true' values might be quite different from each other, and we would not be able to detect it.

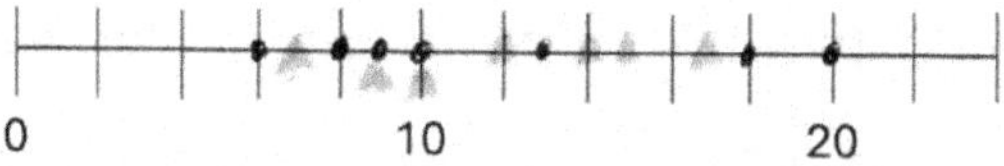

PUZZLE 5: Scenario 2 with data for arsenic sulfate.
Old Chimneys: 8ppm, 9ppm, 6ppm, 20ppm, 10ppm, 13ppm, 18ppm
New Chimneys: 13ppm, 8ppm, 15ppm, 5ppm, 10ppm, 7ppm, 12ppm

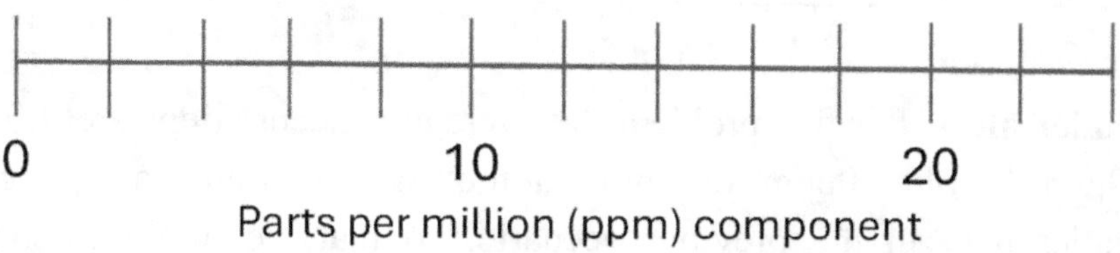

Interpretation goal: Based on our graph, are the values different or not? Frame your interpretation as one of the following: the values are different, the values might be different but we can't be sure, or the values show no indication of being different. The analysis should include an explanation of <u>why do we think that?</u>

Spoiler alert: For this problem, the average concentrations are different (12ppm for old, 10ppm for new), but the variation in the values is large such that the values overlap a lot. Consequently, we might reasonably conclude that the values show no indication of being *meaningfully* different.

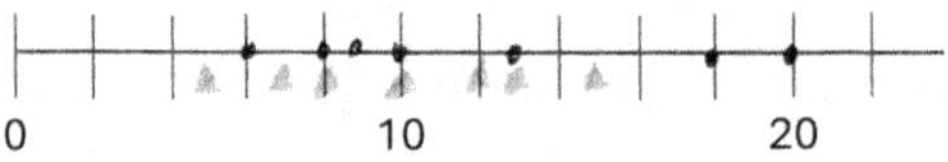

PUZZLE 6: Scenario 3 with data for arsenic sulfate.
Old Chimneys: 12ppm, 12ppm, 10ppm, 14ppm, 10ppm, 13ppm, 13ppm
New Chimneys: 10ppm, 11ppm, 12ppm, 9ppm, 10ppm, 9ppm, 9ppm

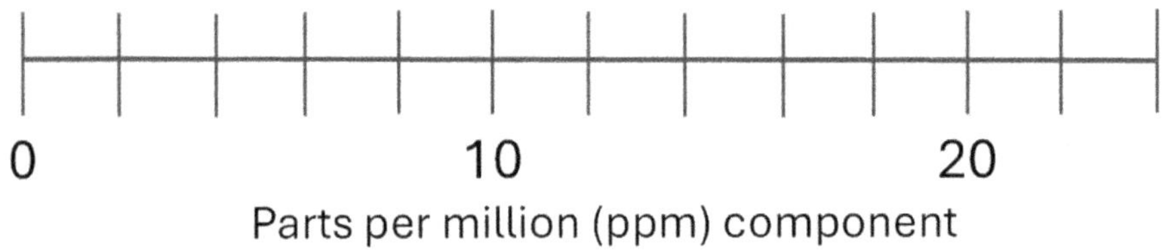

Interpretation goal: Based on our graph, are the values different or not? Frame your interpretation as one of the following: the values are different, the values might be different but we can't be sure, or the values show no indication of being different. The analysis should include an explanation of why do we think that?

Spoiler alert: For this problem the average concentrations are different (12ppm for old, 10ppm for new), but the variation in the values is much smaller than in the previous scenario, so that we might reasonably conclude that the values might be meaningfully different although we can't be sure. We should also recognize that the values do not support an interpretation that the modern chimney contamination can explain the ecological issues in the story, because the old values are actually *higher* than the new values.

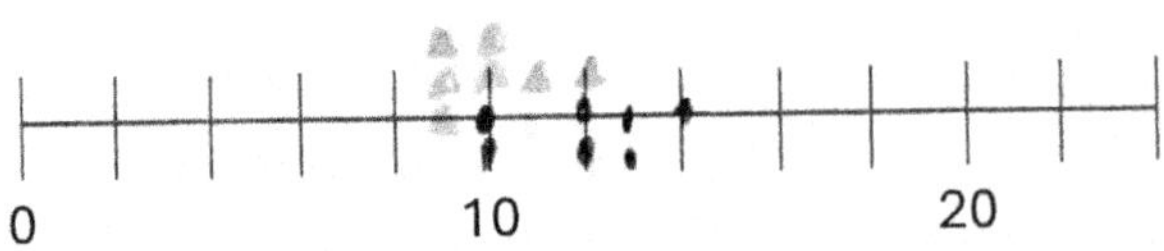

PUZZLE 7: Scenario 4 with data for arsenic sulfate.
Old Chimneys: 10ppm, 10ppm, 11ppm, 10ppm, 10ppm, 10ppm, 9ppm
New Chimneys: 12ppm, 12ppm, 12ppm, 13ppm, 11ppm, 11ppm, 13ppm

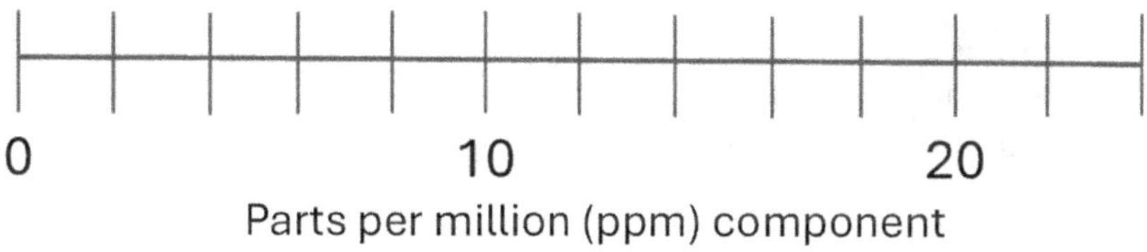

Interpretation goal: Based on our graph, are the values different or not? Frame your interpretation as one of the following: the values are different, the values might be different but we can't be sure, or the values show no indication of being different. The analysis should include an explanation of <u>why do we think that?</u>

Spoiler alert: For this problem, the average concentrations are different (10ppm for old, 12ppm for new), and, although there is still some overlap of values, the maximum variation from the average (+/- 1ppm) is smaller than the difference between them (2ppm) and so we might reasonably conclude that the values are indeed meaningfully different. In this case, the difference might explain the ecological issues in the story, because the old values are lower than the new values.

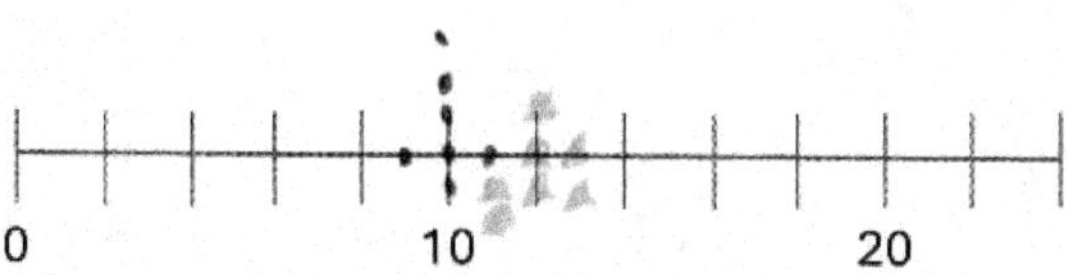

Challenge 3: Plotting and interpretation of multiple chemical components

PUZZLE 8: Scenario 5 with data for arsenic sulfate AND lead sulfate. Often, we can get greater separation of values by considering additional chemical components. Let's consider a situation where we also think about lead sulfate concentration. We can plot two chemical components on an X-Y diagram like that shown below (as before, use different symbols for the old and new chimneys).

Old Chimneys:

Arsenic	**Lead**
10ppm	9ppm
10ppm	9ppm
11ppm	11ppm
10ppm	10ppm
10ppm	11ppm
10ppm	9ppm
9ppm	11ppm

New Chimneys:

Arsenic	**Lead**
12ppm	12ppm
12ppm	13ppm
12ppm	11ppm
13ppm	10ppm
11ppm	13ppm
11ppm	13ppm
13ppm	12ppm

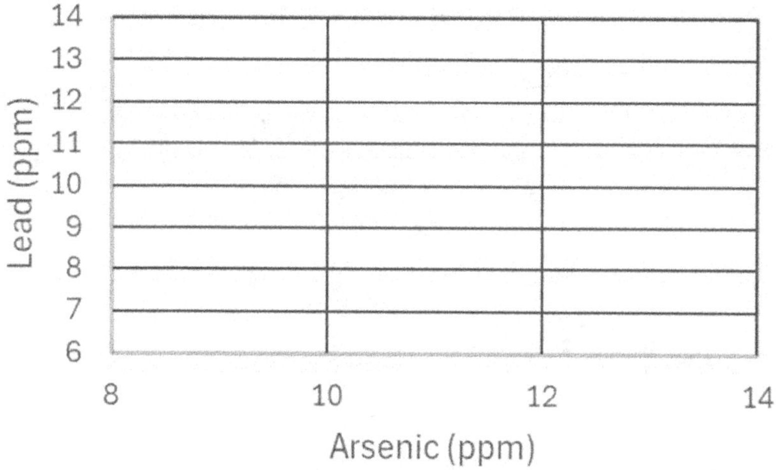

Interpretation goal: Based on our graph, are the values different or not? Frame the analysis as one of the following: the values are different, the values might be different but we can't be sure, or the values show no indication of being different. The analysis should include an explanation of why do we think that?

Spoiler alert: The arsenic data have not changed from Puzzle 7, Scenario 4, and so the conclusions above still apply, but it might be easier to see the separation between the old and new chimneys when plotted this way. Also, we can see that both arsenic and lead are increased in the newer chimneys, adding confidence to our conclusion that there has been an environmental impact in this area between the two time periods. Notice the lack of overlap between the two groupings of points (you can draw non-overlapping ovals around each group).

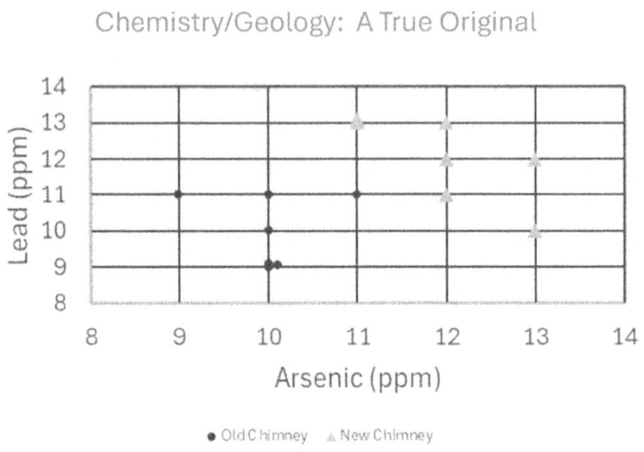

Summary

Without understanding statistical analysis, we cannot understand investigative science in any meaningful way. Statistical thinking is *essential* to analysis of data. However, it is not necessary to understand the mathematics of statistics before engaging in statistical thinking in the analysis of data.

Reflection on Learning

The goal of this book is to practice investigation itself, not to focus on the information and ideas you understood better through investigation. If traditional learning outcomes are not the goal, then what have you learned? Taking some time to consider what you gained from your investigation, reflecting on the practice of science *metacognitively*, can help you hold onto what you have gained. For example: What questions did you ask? What observations and interpretations did you make? Did you *do* modeling, rather than just learning or applying a model?

Real investigation is fluid, creative, and defies easy categorization. Even so, it might be helpful to consider the following categories of investigation offered by the Next Generation Science Standard's science and engineering practices (NGSS Lead States. 2013. Next Generation Science Standards: National Academies Press).

- Asking questions and defining problems
- Developing and using models
- Planning and carrying out investigations
- Analyzing and interpreting data
- Using mathematics and computational thinking
- Constructing explanations and designing solutions
- Engaging in argument from evidence
- Obtaining, evaluating and communicating information

Notice that we have not numbered these practices. They are not steps in a sequential process, but rather components of an integrated and iterative process. Also recognize that this list of practices is **a** way to organize and understand the investigative practice of science, not **the** way. As you go through this reflection, be flexible in thinking about what aspects of your own investigation might be practices of science, because these practices can show up in unexpected ways. Remember, don't think so much about

what you learned or understood, but rather think about what you initiated, did, practiced or experienced.

On a separate sheet of paper, jot down some things that have strengthened your ability to pursue investigative thinking. Give examples of how you engaged in particular, specific practices.

Exploring with you the experience that comes from this particular Learn by Example might provide some guidance as you think about your own practices of science. Let's again consider the practices of science that a reader who works through the offered puzzles might experience.

The reader can find some traditional learning outcomes, in particular important concepts about statistical uncertainty and the nature of chemical analysis. However, learning information or theories is not our goal. Our goal is to practice investigation. In what way does someone who works through the puzzles of this investigation gain experience with the practices of science?

Folks working through the Learn by Example will have a significantly different, and generally less robust, experience with investigative thinking than the person who developed the investigation. For example, questions are presented to them, and they may not engage with asking questions themselves. The investigation has already been planned and carried out for them, and so they may not do any planning or carrying out of their own. They can read through the spoiler alerts to get the "answers," and may not engage in any arguing from evidence or constructing explanations on their own. Of course, if they are approaching the challenges with a properly skeptical mind, they hopefully are questioning the explanations and looking for flaws in them, which is a type of engagement with arguing from evidence in itself. If they are sufficiently skeptical such as to check out aspects of the presented evidence, or consider other arguments that have been presented by others, they may engage with obtaining and evaluating information. If the reader engages with the challenges beyond simply reading through them and looking for the answer, they will participate in additional elements of investigative thinking, such as our examples below.

The most obvious connection to the practices of science is found in **mathematical thinking**. While the puzzles are largely graphical rather than computational, evaluating data in this way, and thinking about the variation in the data and what it reveals about the real-world situation, is a very important aspect of mathematical thinking.

The puzzles also offer important practice with **analyzing and interpreting** numerical data. Plotting of the data on graphs is, in itself, an interpretive practice, but, more importantly, the evaluation of the meaning of the graphical data requires analysis and interpretation of the numbers, not simply "declaring" an apparent answer.

Applying the results of the data to an explanation of its cause, as LaRose begins to do in the story and as is an important aspect of interpreting the data for Puzzles 6, 7, and 8, requires a certain amount of **modeling** to connect the observational data to our real-world interpretation of it. And **arguing from evidence**, and **construction explanations** are implicit aspects of that modeling and interpretation.

Again, perhaps the hardest of the NGSS practices of science and engineering to experience by working through someone else's investigation is "planning and carrying out investigations," which may require more active engagement in doing and developing the investigation. However, to be authentic, your investigation need not engage in all the practices.

Story by L. L. Hill

Nuugyaa

By L. L. Hill

Dealing with a tsunami on her first assignment as number one diver, Hayley not only saves an errant octopus, but the entire Canadian coast from Vancouver to Prince Rupert.

As you can see, the 0600 measurement shows that the side of Mount Georges has continued to bulge, below the slump initially found by RADARSAT." Deputy Minister Siddell tapped on a Power Point slide over a red arrow pointing to disturbed earth, with a caption that read 'increase of almost a meter.'

"The Canadian Geological Survey does expect the shoulder of Mount Georges to slide into Desolation Sound within the next twenty-four hours." Siddell looked around the silent room.

Hayley's mouth felt dry. She wanted to raise her water bottle for a sip but was afraid to move. Yesterday the trees below the cut had still been pointing to the sky. Today they were pointing across the sound—prelude to the collapse of the equivalent of a million transport trucks.

"So today, on station at the Nuugyaa, expect the alarm to sound." Siddell stepped forward and stared at each person in turn. "Get the pumps started as soon as you hear it. Divers, be ready to keep sensors clear."

Someone cleared their throat behind Hayley. She tried to swallow. When Siddell's gaze swept over her again, she was turning the lid with her thumb.

"Any questions?"

The room was quiet.

"Dismiss. Make your way to station ASAP."

Chairs scraped as everyone stood. Hayley sucked water in, delighting in cool relief until an errant elbow bumped the bottle and she dribbled.

"Hayley, have you seen Travis?" Greg asked her as she wiped her mouth.

"No, maybe he's on the boat. He hates briefings."

"Siddell's a stickler for attendance, unless he's in hospital he'll be black listed by the ministry," Greg said as he turned to leave. "Let's go."

Hayley followed, looking for Travis' spikey blond hair. An iridescent red hummingbird hovering over a blossom outside the door distracted her. She wondered if birds worried about mountains falling into the sea. Not like an earthquake, she thought.

With her next step, she was on the floating dock which rebounded as dive crews tromped to the 'not since Dunkirk' multi-hued assortment of vessels which had called the improvised town of Haskins Inlet home for four months. Resolve pulsed with every step.

"Hayley's number one diver today, Tom," said Greg as he opened the engine cover of their converted fishing boat.

Tom looked at her and turned away, his queue shaking like a snake. "We getting a spare?"

"Olaf is on his way."

Wondering when Greg had called Olaf, Hayley began routine checks of her dive gear. She felt a quiver in her legs and drank more water.

"Have you ever been number one, Hayley?" Tom asked.

"No." She looked into his black eyes. "No one's ever tried to mitigate a tsunami created by a massive rock slide either."

He snorted and turned back to check his gear.

Hayley raised a hand and dropped it to her side. "I know that you have twice as many hours, feel free to make suggestions."

"I don't think that a number one diver should need any suggestions," Tom said.

Hayley went crimson under her freckles.

“Enough Tom, you’ve been out of the water for eighteen months and you were never perfect either. Hayley, ignore Tom,” Greg said as he walked to the cabin.

Twin diesels rumbled as Hayley inspected air hoses. No one had said why Tom had taken time off diving and Hayley did not mind not knowing. She did mind being told to ignore someone. Sooner the mountain fell the better, she thought as she checked her helmet.

As other boats began to depart, Olaf rushed down the dock with his gear bag. He nodded at Tom before he talked to Greg in the cabin. He came out and circled the boat, casting off lines. Then he slapped Tom on the back and began a private conversation with him as the boat slipped away from its mooring.

Hayley grimaced. Thinking that if she were Tom, she might be upset with the situation, she decided to keep her focus on her gear.

Greg had the snub bow pointing after the flotilla heading to the nine-kilometer-long network of pipes and pumps known as Nuugyaa—Haida for octopus—that would dam Desolation Sound. Steep green fjord flanks plunged into blue water. On tapered cable, floating below silver helium-filled airships a kilometer above, were the wind turbines that powered the array. Offset pairs of blades ten meters apart provided stability and assisted lift from their maple seed-like design. Although the siphon effect reduced the power required, these elegant machines could power the entire West Coast. If anything failed, sections separated and floated down.

Bald eagles watched the fleet as it passed beneath their evergreen perch where Haskins Inlet met Desolation Sound. Once, Hayley had asked Tom about myths of interactions between eagles and humans. He had glared at her and walked away. Hayley wondered how she would feel if someone asked her to define Christmas or Easter. She did not think that she would turn away from them. After this job, she would find another company to dive for. Greg was a competent supervisor, but Hayley wanted to feel part of a team and not a solo act.

Hayley brushed her socks off and slid her long legs into her dry suit. Then she stood and popped her suspenders on her shoulders before powdering her seals.

Words blew out from the cabin, "...too reliant," and wafted across the deck. From the corner of her eye, Hayley saw Tom stiffen and look at her. She sat down and stretched her legs.

'Too reliant?' She never discussed her personal life, finances, or relationships. She never asked for that information from others, unless religious beliefs were counted as personal. While Travis had yacked a kelp bed about his life, neither Greg nor Tom did. She had asked for opinions on gear and listened. No one had asked her why she had chosen her gear.

"Ready to go? Inspected the hoses?" Greg yelled out the window. "Slack tide now."

She made an 'okay' sign and watched the pump barges and yellow buoys scattered over Desolation Sound grow larger. Hayley hoped that the lattice of pipe beneath them worked as planned. If one of the sensors monitoring water movement failed, flow of the section was affected. Vancouver to Prince Rupert was a lot of coast to evacuate. The narrow passages would exaggerate any tidal wave just as they did the daily tides. Responsibility weighed on her like fathoms of sea water.

Beyond her vision was the green clad side of Mount Georges. It was hard to imagine a mountain sliced in half, no trees left and only rocky soil washing away in the rain. No attempt to deflect the massive landslide could be made. It was anticipated that trees a half kilometer up the side of Mount Bartholomew across the sound would be flushed away. Live feed cameras would record the event. Hayley imagined a wall of water, rock, and trees rushing towards her. She shivered. Hard to imagine that on a sunny day.

Greg throttled back as they approached their mid-zone station that serviced thirty pipes on Nuugyaa. The tug waiting there slipped its mooring and drifted back. Olaf went forward to snag the buoy line. Hayley slipped her arms in and popped her head through her neck seal.

Her ears came through to the alarm wailing. Hayley jumped and looked up the sound as the siren echoed off the mountainsides. Greg was yelling at Olaf who had missed the mooring and dropped the boat hook.

Hayley lunged over the rail, using her long reach to grab the hook as it floated by. Grunting, she hauled herself up and ran forward past Tom's wide eyes.

"Get the buoy!" Greg shouted.

Around them, pumps on rafts rumbled, priming the siphon of water in one end and pushing it out the other to disrupt the energy of the tsunami. Balanced below pump, each thirty-meter section of neutrally buoyant two-meter diameter pipe ended in an arrow shaped fan that could siphon or pump. As the tsunami rolled over, the siphon in flow would begin the process of disrupting the energy, which would be continued by pumping water out.

Hayley hooked the buoy line as Greg goosed the boat forward and Olaf fell overboard.

Distant thunder of the avalanche nearly drowned the siren and the pumps as Hayley yelled out of reflex, “Man overboard!”

Hayley released the line and reached for Olaf. He seized the hook and she walked him back to the stern where Tom lowered the ladder for him. Hayley dashed past red-faced Greg and caught the line again. She looped the mooring line through the eye, tied it off and raised her hands.

As the boat drifted back, Hayley looked over the confused sea, almost a million vortexes and fountains speckling the water amidst boats struggling to moor. On Ajax Island, trees shivered as the Earth shook.

“I need you dressed, Hayley. A sensor is malfunctioning,” Greg yelled into her ear.

“Okay, dressing,” Hayley shouted, heading to the stern.

“Diver dressing,” Greg called to Olaf as he stripped in the cabin.

“Zip,” Hayley demanded of Tom, standing in a cross.

She felt Tom tuck her underwear beneath the zipper as the brass teeth closed. He patted her shoulder and watched as Hayley pulled the last centimeter of zipper shut. Kneeling, she broke her neck seal to vent excess air, glad that she could move and tremble at the same time.

Slipping into her harness, a shoulder strap twisted. Hayley felt sweat bead on her lip as she reached back to straighten it.

Tom grabbed her shoulder as he flipped it. Hayley felt the compressor start. Then Tom pulled her arms into her BC. She felt her inflator hose click into place.

Hayley dropped her fins to the deck, letting Tom push her feet in and pull the straps tight. With fingertips, she checked her water bottle and tool strap. Then she put her liner gloves on. She let Tom snug her dry gloves on over her wrist seals.

"Helmet!" Hayley yelled.

Dry air filled the helmet when she inhaled. Most of the sound of the demise of Mount Georges was sealed out.

"Comm check," Olaf said.

"Comm check, okay," Hayley replied.

"Diver ready?" Olaf asked as Tom hung her umbilical through a carabiner on her harness.

Hayley closed her eyes and took another deep breath as she made another mental check of her equipment.

"Diver ready," she replied as she stood.

Legs spread, she waddled to the ladder. Glad to feel the water take her weight with each step, sweat trickled down her back as she stepped down.

"Diver in the water," Hayley said as she sank beneath wavelets crowned with sparkling gems.

"Diver okay?"

"Diver okay," Hayley replied as she equalised her ears.

"How's viz?"

"Three, four metres."

An orange Lion's Mane jellyfish brushed her shoulder in the hazy water. Hayley wondered if Greg and Olaf could hear her heart hammering as she drifted down to the Nuugyaa suspended ten metres below the surface. The power line to the pump disappeared like a thread off a hem though it dangled to the bottom eighty meters below Nuugyaa.

"Which sensor?" Hayley asked.

"Middle one shows offline."

Hayley finned over to the elbow that rose to the fan. She touched the algae on the turned off composite pipe and swam forward. Likely kelp wrapped around the sensor orifice, she thought. She could just see the shadows of pipes on either side of her as she kicked. Mentally she rehearsed the procedure if she got sucked in: exhale, cut lines, switch to SCUBA, steady exhale out the rear orifice. A quick trip she hoped never to make. She added air to her dry suit and shook out a squeeze in one leg before finning forward.

She wondered what Mount Georges looked like now. The scene was likely already getting replayed for news viewers trying to see energy waves ricocheting through the Inland Passage.

"Diver okay?"

"Diver okay. What's it look like up there?"

"Like Mount Georges had a very bad shave. Chopper pilot says the wave beat the estimated height. Energy waves reflecting will take a while to dissipate."

"Understood."

Hayley flicked off a starfish that had found a temporary home. She wondered how many starfish it would take to plug the sensor orifice.

A wall of water hit her. She slipped back and then held tight to a rung on the pipe. The tide had ebbed right on schedule. Not that it would change the tsunami's power.

"Diver okay. Tide is ebbing."

"Hope you find the problem."

"I must be near," Hayley said pulling herself along rungs placed for diver service.

Each breath rasped in and bubbled out. Just ahead she could see an anchor line of the intake. Being wiped from the hose by the current did not occur to her.

"At the anchor point."

"Let me know when you are at the sensor."

"Understood."

Hayley hung on, feeling like a flag blown from its pole. Pinching her knees around a rung, she pulled a Jon line out of a BC pocket and curled it around the next metal loop. Then using a knee to press down, she hauled herself up to the sensor like a lumberjack.

"Diver at the sensor." Hayley struggled to hold on and feel inside. "I feel a pulpy mass."

"Is it kelp?"

"No," Hayley said as she cinched the Jon line tight in one hand to pull herself up. "It doesn't feel like kelp."

Hayley clenched the Jon line and finned hard to hold her place. She shuffled up the line, swaying like a kite on a string. Her heavy breathing would be clear on the comm, she thought. She had heard worse. One guy had lost his beer in his helmet.

Now she could see inside the flared lip of the sensor orifice. Hayley reached out and a tentacle tip touched her finger. She started and released a bellyful of bubbles.

“Diver okay?”

“Diver okay. Problem is an octopus.”

Hayley looked into slit pupils, noting the horns were up as the octopus blushed from brown to red. She stroked the small suction cups and imagined all of them holding the creature in place.

“Tom says that it is appropriate that Nuugyaa find a home in Nuugyaa.”

“No kidding. Even with all the racket, I’m going to have to persuade Nuugyaa to find a new home.” Without hurting it, she added to herself.

Hayley pulled on the arm and watched the octopus shrink back into its shelter. She had heard that bleach flushed them out but had never tried it herself. Tap water had some chlorine in it. After taking a sip from her water bottle, she slid the hose from under her helmet. Needing a third hand, she pinched the Jon line between her legs. Feeling that she could go upside down, she squirted her bottle at the Nuugyaa. In a frantic flare of tentacles, the Nuugyaa erupted out, jetting ink and swimming backwards.

“Blockage removed. Hope that Tom will let me know the correct apology to make to Nuugyaa.”

Greg coughed. “He will, Diver.”

“Diver returning.”

“Diver returning,” echoed the surface.

Hayley looked around. If she ascended, her umbilical might catch on a buoy line. She opted to descend and seek the lee of the pipe from the current.

“Descending to main pipe,” Hayley said as she tried to float down the pipe. Her shoulders would be sore tomorrow, she thought.

“Understood Diver.”

“Diver okay?”

“Diver okay.” Like they could not see, she thought.

“Float up, Diver.”

“Understood. The dirty water coming will kill a lot of animals.” She hoped the octopus would survive.

“Thanks to Nuugyaa, most of them won’t be human.”

“True. Diver ascending.” Corny but true, Hayley thought as she drifted free. “Clear of Nuugyaa.”

“We’ll pull you in, Diver.”

Relaxed, Hayley raised her hands over her head. At the surface, the sea looked like a boiling sewage lagoon. She rolled over and finned to the stern ladder. Three men smiled as the pipe erupted into service. Hayley hoped that they had something to smile about.

“Good job, Diver,” Greg said, as the alarm changed to a rapid cheep. “Let’s get you out.”

Spurred by the increased urgency in the alarm, Hayley climbed the ladder with stiff knees as Tom pulled on her BC.

“Diver on board,” said Greg.

With the crew, Hayley looked past the bow of the boat at the peak of a massive wave, studded with trees, that was rushing towards them from almost five kilometers away. It was like standing on a highway and watching sixteen lanes of eighteen-wheelers approach. Cavities appeared in the face as the siphons pulled water away while fountains of ejected water created chop. Each kilometer of Nuugyaa battered the wave to smaller pieces. Boats bobbed on the now jagged crest and the massive pipes pivoted as the diminished wave rolled on.

“The anchors are holding,” Hayley said, glad that no pipes had broken free.

“Yeah, it’s breaking up,” said Olaf.

As the wave rolled under them, the boat jerked on its mooring lines. Hayley watched a spruce pivot by, a plumose anemone protruding like a posy from its spines, brushing the bow of the boat. Spikes of steep waves tossed the moored flotilla behind them as the tsunami dissipated.

Hayley sagged to the dressing bench. Tom helped lift her helmet off.

"Looks like Nuugyaa took the stuffing out of the number one diver as well as the tsunami," Tom observed. "You ready to go if another sensor goes?"

Hayley watched over the stern as the departing shift of boats tossed on seas smaller than a boat wake. She smiled.

"If another sensor fails, I'll back up the number two diver."

***

L. L. Hill is currently resident in northern Canada; writing is a hobby competing with photography and planting wildflowers for her spare time. A novel is available on Amazon and features Laura's interest in crystals, ley lines, and a scientific approach to the paranormal. A second novel, "Road Rogues," is near publication. Crystal Gaze-Key West: Hill, Laura Lee: 9798871563823: Books - Amazon.ca.

Art Credit: Nuugyaa artwork by Erin Colson.

"Nuugyaa" first appeared online in *Issues in Earth Science*, Issue 7, July 2017.

Chase Your Own Rabbits!

Question: Find something in the story that you wonder about, or want to challenge, or would like to explore more fully, and ask a scientifically-investigable question about it. As you consider things you wonder about, remember that not all questions can be investigated scientifically. For example, for the story *Nuugyaa,* a question like "would a non-volcanic mountainside actually bulge prior to collapse like is stated in the story?" might simply lead to an online search for the "answer" and would not be investigative. Alternatively, you might narrow the question to something like "does the volume of a material increase just prior to failure such as might be due to formation of small fractures within the material?" This is a question that could be addressed experimentally with analogs, even wooden pencils! Other questions, some of which could be addressed through investigation and some not, might grow from the statement in the story that the "energy waves reflect" off the surrounding topography. Try to find a question that can be investigated scientifically with resources that you have, or can gather, in the classroom.

The question we, Russ and Mary, address in our Learn by Example for the story *Nuugyaa* is the following:

"Are the movements of mass and energy implied by the descriptions in Nuugyaa reasonable or even possible?"

You can choose this question as a seed-question for your own investigation or come up with your own rabbit to chase!

Feel free to write down your question here:

Vague Notion: As you prepare to plan your investigation, first get a vague notion of what you plan to do—including guidelines on whether it will

involve experimentation, data-interpretation, modeling, arguing from evidence or some other practice of science.

The vague notion that we started from for our Learn by Example for this story is the following:

We can approach the problem of the reasonableness of the mass and energy movements like engineers, considering the constraints of mass balance, energy balance, and logistics. With these considerations, we can go through a series of calculations that give us an understanding of the amounts of energy and mass involved in the collapse of the mountain and what would be required to mitigate the ensuing tsunami.

If you chose to go with our question for your own investigation, you can also choose our starting vague notion, or chase your own rabbit.

Feel free to write down your vague notion here:

Detailed Planning: Once you have your question and a vague notion in place, you can begin to fill in the nuances, complexities, and details of your plan. There is no recipe for your plan or a predetermined set of steps to follow, because your plan ***is*** the recipe or set of steps that you are creating for this particular investigation! Sketch out your investigation, anticipating as best you can what challenges you will encounter, and include as much detail as you are able (realizing that your investigative plan may not survive the first results of actually doing it.) Be sure to think about what background information you might need and how to get it, techniques you may need to research, the overall sequence of things you

will need to do, and all other aspects of your investigation! You will likely need several pages to document your plan.

Follow a Question

Asking a question, choosing a vague notion of how to pursue that question, and then developing a detailed plan to pursue it is just the start of investigating. You actually have to do your investigation! During the course of doing your investigation, you will find out what works and what doesn't work. You may realize that you need more information, or different experimental tools, or even that your initial working model was so far off that your plan for investigation quickly falls apart. Authentic investigation involves iterative thinking during which your original experimental plan is revised, your initial interpretations updated, and your working model revisited. All of this is the whole point and practice of authentic investigation!

Keep careful notes on your experimental methods, interpretations, and models as well as your revisions as you work through your investigation.

Learn by Example (for Nuugyaa)

Our Starting Question: "Are the movements of mass and energy implied by the descriptions in Nuugyaa reasonable or even possible?"

Our Starting Notion: We can approach the problem of the reasonableness of the mass and energy movements like engineers, considering the constraints of mass balance, energy balance, and logistics. With these considerations, we can go through a series of calculations that give us an understanding of the amounts of energy and mass involved in the collapse of the mountain and what would be required to mitigate the ensuing tsunami.

Introduction

Nuugyaa does not lack for grand-scale events and ambitions! Check out the following excerpts from the story:

"The Canadian Geological Survey does expect the shoulder of Mount Georges to slide into Desolation Sound within the next twenty-four hours." Siddell looked around the silent room.

... "Have you ever been number one, Hayley?" Tom asked.—"No." She looked into his black eyes. "No one's ever tried to mitigate a tsunami created by a massive rock slide either."

Sadly, grand geological events of this scale have a notorious record in fiction, ranging from highly improbable to outright impossible—just check out any disaster movie with an earth science event in the title: earthquakes, volcanoes, tornadoes, climate change, take your pick.

So, do the events in *Nuugyaa* make any kind of sense, and how can we tell?

Nuugyaa includes a number of elements of interest in earth science, including a tsunami, a slump, and a landslide. However, the main science element embedded in the story is the idea of mass and energy balance.

Mass balance and energy balance are the ideas that matter and energy are neither created nor destroyed and that we can keep track of the amount of mass and energy as they move from place to place. Understanding mass and energy balance is important not only for understanding the science of the origin of the tsunami, but for understanding the mitigation system put in place to save the coast of Canada from the tsunami.

Mass and energy balance are, in fact, the key ideas in all of earth system science. Understanding global climate and climate change, earth carbon balance, pollutant migration, global precipitation patterns, etc., are all problems in mass and energy balance. To some extent, engineering involves applying science ideas—like mass and energy balance—to solving problems. Many problems—like the problem of mitigating the tsunami in the story—require thinking in these terms.

Thus, *Nuugyaa* provides a great opportunity to both engage in thinking about mass and energy balance and in applying that science thinking to a real problem—a perfect application of STEM thinking. From an engineering perspective, there are at least three types of constraints on the Nuugyaa tsunami mitigation system.

- Energy balance. Does the amount of energy delivered by a collapsing mountain justify the size of the wave—that is, is the energy of the collapsing mountain equal to or greater than the energy of the tsunami? Is the energy available to the protagonists sufficient to counteract the energy of the wave?
- Mass balance: Is the infrastructure put in place to stop the wave adequate to move the volume of water implied by the height of the wave?
- Logistical constraints: Is there some realistic method for moving water such that the energy of the wave might be dissipated? Is there realistic reaction time for the protagonists to actually take any meaningful action?

Let's consider some of these constraints in an effort to understand how systems work, both in nature and in engineering problems.

Remember, the following puzzles and challenges are not intended to be systematic curricular steps through which students of science work toward defined learning outcomes. Rather, they are intended as aspirational examples of investigative thinking, appropriate at a variety of levels—some middle school, some high school, some for the expert teacher—to inspire investigative thoughts and increase understanding of what investigative thinking is all about. Find one or more that is right for you!

Challenge 1—Energy of the original landslide

Trying to figure out *how* to calculate the energy of the landslide is probably a better starting point than simply plugging numbers into an equation that someone gives you—after all, learning how to push buttons on a computer or calculator is not really a very valuable goal of science investigation. However, figuring out what equations to use and how to use them is a harder challenge than simply plugging numbers in! We need some ideas to get started.

To help us get started, let's review some ideas from physics.

How do we determine potential energy or kinetic energy of a suspended or falling object?

Kinetic energy = ½ x mass x velocity squared.

From this relationship, we can get the energy that an object would have if it fell a given distance in Earth's gravity field by calculating how fast the object will be travelling after falling that given distance.

Notice that if one were to use this equation to determine energy, energy being the goal of our challenge, they would need to know something about mass and velocity. So, let's think about velocity.

Velocity = square root of (2 x height x acceleration of gravity).

Acceleration of gravity on Earth is about 9.8 meters per second squared. Notice that to determine velocity using this equation, one needs to know something about the distance of fall.

Alternatively, we might start with the equation for potential energy.

Potential energy = mass x height x acceleration of gravity.

Notice that, using this equation to determine energy, we again need to know information about mass and distance of fall.

The concept at the heart of this challenge is that the potential energy at the start of the landslide must equal the total energy dispersed by the landslide, one key idea of energy balance.

Let's approach the problem in a series of steps. To get to energy, we need to know mass. To figure mass, we need to know volume.

Puzzle 1: Estimate the volume of the rock of the landslide.

From the story, the volume of rock in the landslide is "the equivalent of a million transport trucks." What is the mass of the landslide?

Remember, this is an investigation, not a calculation. It's not like a class assignment where someone else has put together for you all the information that you need! Figure it out! Make approximations where you can't get an exact number.

Spoiler Alert: Trucks come in quite a few different sizes, so we can't get an exact number from this. However, an online search on "volume of transport truck" yields a mid-value of around 30,000 liters for large tanker trucks. For comparison, a typical gravel dump truck holds a little over 10,000 liters. Anyway, 30,000 liters gives us a ball-park number. A million transport trucks then suggest a volume of 1,000,000 x 30,000 = 30 billion liters.

PUZZLE 2: What is the mass of the rock?

Our equation for energy requires us to use mass, not volume. How do we estimate the mass of 30 billion liters of rock?

Spoiler Alert: An internet search on "density of rock" gives us a typical range from 2 to 2.9 grams per cubic centimeter (we are going to use a value of 2.6). There are 1000 cubic centimeters in each liter.

(30 billion liters) x 1000 cubic centimeters per liter x 2.6 grams per cubic centimeter = 7.8×10^{13} grams = 7.8×10^{10} kilograms

Be sure to take note of the units and unit analysis to make sure you are doing the calculation correctly. If a calculation is done correctly, the units have to cancel out to yield the correct final units. In this case, the units of liters are divided by liters, cubic centimeters are divided by cubic centimeters, leaving grams, which can be converted to kilograms.

PUZZLE 3: So, what is the energy of the landslide?
From the story, the rock falls about a kilometer, or 1000 meters. The mass is calculated above from the estimate of a million transport trucks. What is the energy?

Spoiler Alert: We have the mass, the distance of fall, and the acceleration of gravity, but not the velocity of the landslide, so the best equation to use might be

Potential energy = mass x height x acceleration of gravity.

7.8×10^{10} kilograms x 1000 meters x 9.8 meters per second squared =
7.6×10^{14} Kilogram-meter squared per second squared (kgm^2/s^2) =
7.6×10^{14} joules. (1 joule = 1 $kg\text{-}m^2/s^2$ = the standard SI unit of energy).

It might be interesting to compare the energy of the landslide to other types of energy to get a feel for what this number means. For example, how much energy does burning a gallon of gasoline release? An online search shows that a gallon of gasoline has about 120 megajoules of energy. So, we can calculate that the landslide releases the equivalent of

7.6×10^{14} joules/ 120×10^{6} joules per gallon = 6.3 million gallons of gasoline

Challenge 2—Does the energy of the landslide account for the size of the tsunami?

PUZZLE 4: Figure out the energy of the tsunami. The calculation of the energy of the tsunami (a calculation only for the brave!) is less straightforward than the landslide energy calculation because the speed, height, length of the wave crest and wavelength of the wave are complex functions of the water depth and surrounding landscape. However, engineering often involves making a 'best estimate' of values and so an exercise like this can be a valuable experience in how engineering problem-solving proceeds (but not something to be accomplished in a couple of class periods!). The story gives us an estimate of the wave height, since it went more than 500 meters up the side of Mount Bartholomew, however, we need to get creative in estimating other aspects of the tsunami. Give it a whirl if you have plenty of time and a courageous spirit!

Spoiler Alert: An online search yields a paper that gives an equation for wave power (although the equation is valid for deep water, so it may only approximate the energy of the shallow water tsunami of the story). From the reference—Amir Vosough (2011) Wave Energy, International Journal of Multidisciplinary Sciences and Engineering V 2.—we can get the following:

Power = 0.5 (kilowatts/m^3-s) x length of wave crest x height of wave squared x period of wave

Note: Energy = power x time.

Period of the wave is the time for one wavelength to pass.

We estimate the wave height is 500 meters based on how high up the side of Mount Bartholomew it went. We can make some guesstimates for our imaginary tsunami of the story and the scale of the waterway in which it occurs—suppose the wavelength of the wave is about 500 meters, comparable to the wave height, and the length of the wave crest is about a kilometer (1000 meters) (perhaps reasonable for a wave arising from a point-source landslide into the constrained confines of one of the channels along the Strait of Georgia in British Columbia) and that the wave travels about 160 kilometers per hour (this is comparable to the speed of the real-world tsunami at Lituya Bay in Alaska in 1958). We get

$$\text{Power} = 0.5\ \text{kW/m}^3\text{-s} \ \text{x}\ 1000\text{m x } 250{,}000\text{m}^2 \text{ x } 11.25 \text{ seconds}$$

$= 1.4\text{x}10^9$ kWatts

Or converting to energy (power x time) = $1.4\text{x}10^9$ kWatts x duration of wave (that is, the time for the single wave to pass, 11.25 seconds) = 4.4×10^6 kilowatt-hours.

Using a handy internet converter gives us 1.6×10^{13} joules of energy in the tsunami.

PUZZLE 5: Can the energy of the falling rock account for amount of energy present in the tsunami? What percentage of the total energy of the landslide is accounted for by the energy in the tsunami? Is it the same, or more or less?

Spoiler Alert: Yes, there is probably enough energy in the landslide to account for the tsunami. We calculated the energy in the landslide to be 7.6×10^{14} joules. This is an order of magnitude more energy than the 1.6×10^{13} joules in the tsunami.

Puzzle 6: Where does the rest of the energy go (an energy-balance consideration)? Brainstorm some thoughts on what happened to the rest of the energy, the difference between the 7.6 x 10^{14} joules of the landslide and the 1.6 x 10^{13} joules in the tsunami. The wave energy is only about 2.1% of the energy from the landslide (1.6 x 10^{13} joules/7.6 x 10^{14} joules). One order of magnitude is a BIG deal: Remembering our idea of energy balance, where did the rest of the energy go?

Spoiler Alert: Can you hear the landslide? Might energy be going into making sound waves?

Can you feel the vibrations of the landslide through the ground? Might energy be going into making seismic waves?

What might happen to the rest of the energy? Throwing rocks around? Creating a gust of wind? Friction, within the landslide, between tsunami and ocean floor, or within the water column, converting the energy to heat?

Challenge 3: Could a system like the Nuugyaa mitigation system really dissipate this much energy?

Puzzle 7: If the tsunami has 1.6 x 10^{13} joules (4.4 x 10^{6} kilowatt-hours) energy, it will take at least that much energy to dissipate it. According to the story, the power is supplied by wind turbines. How many wind turbines would it take to supply that much energy, say over a period of 3.4 minutes—the time it takes the tsunami to travel about 9 kilometers (that's the length of the Nuugyaa network as reported by the story)?

Spoiler Alert: Power production depends on wind speed and height of the turbine as well as its capacity, however, an online search indicates that typical wind turbines today generate on the order of 500 kilowatts power on average. Remember that energy = power x time. For power used over the 3.4 minutes it takes the wave to move through the Nuugyaa system, that would correspond to an energy of about 500 kwatts x 0.057 hours = 29 kilowatt-hours energy or 1.04×10^8 joules. Thus, we can calculate that the Nuugyaa system would require about $.6 \times 10^{13}$ joules/1.04×10^8 joules = 580,000 such turbines on an average day (hopefully the tsunami wouldn't happen during a lull in the wind!).

Challenge 4: How much water does the Nuugyaa system have to move in the 3.4 minutes of the wave's passage (a mass-balance problem)

PUZZLE 8: A simplified way of approximating the volume of water in the wave is to consider the wave to be triangular shaped in cross-section (ok, so that isn't such a great approximation—if you want to use a sine wave function, go for it!)

Spoiler Alert: With the triangular assumption, the volume of water will be the cross-sectional area of the triangle times the length of the wave. With a wave height of 500 meters, a wavelength of 500 meters, and a length of crest of 1000 meters (same numbers as used above), we get a wave water volume of

0.5 x 500m x 500m x 1000m =
125,000,000 cubic meters of water = 125 billion liters

PUZZLE 9: That's a lot of water to move in 3.4 minutes. The story features a network of 2-meter-diameter pipes in sections of which the mid-section has 30 pipes. What is the total rate of water movement required to move 125 billion liters of water in 3.4 minutes (discharge can be thought of as volume per unit time, or liters per second for example)?

Spoiler Alert: Rate of water movement = 125 billion liters / 204 seconds. Those pipes have to move water at a rate of 613 million liters per second.

PUZZLE 10: How many pipes would that reasonably take? The pipes in the story have a 2-meter diameter.

Spoiler Alert: With a pipe diameter of 2 meters (as reported in the story) and a reasonable rate of flow (say a maximum of 5 meters per second extrapolated from an internet search on maximum flow rate in pipe by size).

discharge = cross-sectional area of pipe x velocity
Cross sectional area = πr^2
Yielding discharge = $\pi \times 1m^2 \times 5m/s = 15.7m^3/s$ maximum

This corresponds to 3200 m^3 for the entire 3.4 minutes per pipe.
(15.7 m^3/sec x 204 seconds)

To transport 125 million cubic meters of water in 3.4 minutes would take a minimum of
125,000,000 m^3 /3200 m^3/pipe = 39,000 pipes of the size described in the story.

Benefits of STEM Thinking

There is a tendency for us to adopt a type of magical thinking concerning how technology can solve our problems, perhaps because so much of our modern technology does come to us seemingly like magic. Hopefully this exercise can give an understanding not only of how to approach an engineering problem like a scientist or engineer, but a sense of why protecting ourselves from natural disasters is so difficult. We have to balance costs and benefits. In the story *Nuugyaa*, for things to work, people would need to build 580,000 wind turbines and a system of nearly 40,000 nine-kilometer-long pipes and pumps to protect from a one-time event that we don't know for sure when (or if) will happen. Wow.

Reflection on Learning

<u>The goal of this book is to practice investigation itself</u>, not to focus on the information and ideas you understood better through investigation. If traditional learning outcomes are not the goal, then what have you learned? Taking some time to consider what you gained from your investigation, reflecting on the practice of science *metacognitively*, can help you hold onto what you have gained. For example: What questions did you ask? What observations and interpretations did you make? Did you *do* modeling, rather than just learning or applying a model?

Real investigation is fluid, creative, and defies easy categorization. Even so, it might be helpful to consider the following categories of investigation offered by the Next Generation Science Standard's science and engineering practices (NGSS Lead States. 2013. Next Generation Science Standards: National Academies Press).

- Asking questions and defining problems
- Developing and using models
- Planning and carrying out investigations
- Analyzing and interpreting data
- Using mathematics and computational thinking
- Constructing explanations and designing solutions
- Engaging in argument from evidence
- Obtaining, evaluating and communicating information

Notice that we have not numbered these practices. They are not steps in a sequential process, but rather components of an integrated and iterative process. Also recognize that this list of practices is **a** way to organize and understand the investigative practice of science, not **the** way. As you go through this reflection, be flexible in thinking about what aspects of your own investigation might be practices of science, because these practices can show up in unexpected ways. Remember, don't think so much about

what you learned or understood, but rather think about what you initiated, did, practiced or experienced.

On a separate sheet of paper, jot down some things that have strengthened your ability to pursue investigative thinking. Give examples of how you engaged in particular, specific practices.

Exploring with you the experience that we, the editors, had in developing our Learn by Example section might provide some guidance as you think about your own practices of science. There are some traditional learning outcomes in these exercises, for example, consideration of units, power versus energy, volume versus mass, application of common physics equations, examination of energy balance, and others. However, learning information or theories was not our goal. Our goal was to practice investigation. The practice of science that we highlight in the Table of Contents for this Learn by Example is mathematical thinking. In what way did we, the editors who created and implemented this investigation, and we, the readers who work through it, gain experience with this practice?

As you can probably imagine just by glancing through the exercises in our Learn by Example, we, the editors, had to go through a pretty significant learning curve in developing these challenges. We had to do extensive online research to come up with reasonable equations to use, find information about rock density, typical energy production of wind generators, flow capacity of pipes of various sizes, and many other items, and then make reasonable choices about whether various approximations were reasonable or not. All of these are important elements of **obtaining and evaluating information**. Then, the practice of setting up the calculations, deciding what data we needed, what equations to use, what units to work in, what data values to compare, etc.—the heart of

mathematical thinking—ranged from fairly straightforward to quite complex. Thus, developing this investigation gave us a significant experience with these particular practices of science.

For the readers of this book, their experience with mathematical thinking might be less robust. Just reading through calculations, or even working through them, is not the intention of mathematical thinking as defined in the NGSS practices of science and engineering (See NGSS Lead States. 2013. Next Generation Science Standards: National Academies Press, Appendix F, Mathematical and Computational Thinking).

Mathematical thinking requires engagement with the *development* of mathematical models, or the construction of *steps, processes, or approaches* to a mathematical solution. We invite that type of participation with most of the challenges in this Learn by Example. Hopefully many readers find some that they can challenge themselves to figure out before reading through the Spoiler Alert.

However, the math is often sufficiently challenging that many readers will work through the equations and calculations as presented in the Spoiler Alert without first attempting their own mathematical derivations. This is an OK place to start. It is helpful to work through steps already developed by someone else as a way to begin to understand how to figure out how to develop a mathematical solution beyond plugging numbers into an existing solution, which can be difficult in itself! We, the editors, believe that working through solutions developed by someone else, especially challenging ones such as some of the ones presented here, can engage in authentic **mathematical thinking**.

Story by Justin Short

Breaking the Ice

By Justin Short

The Ringlets, newly settled in Antarctica from the rings of Saturn, are ready to go to war but refuse to negotiate with anyone over 17. Can young Gwen figure out what made them so angry soon enough to prevent an interspecies war?

The icebergs were orange and purple tonight, and they were seriously gorgeous. Like, cotton-candy-fingers-against-a-perfect-sunset gorgeous.

“You see it, Gwen?”

Lon pointed over the ship’s railing, and I tried to follow his finger.

There it was. A gigantic iceberg in the middle of all the others. Jagged and hilly. It shot from the dark water like a lonely volcano.

This whole Antarctic thing was tough sometimes. Mainly the temperature part. But it was worth it for views like this.

Did I mention the cold? I'm talking true, lung-freezing cold. And trust me, I wore layers. Lots of 'em. Three shirts, long johns, flannel jeans, thick boots. A puffy coat that zipped all the way to my chin and made me look like a loofah with feet.

"All right," Lon said. "Let's get this over with." He swung his leg over the railing and dropped into the tiny ship-to-shore boat. I followed him in. The sailor on duty lowered the boat down with the crank and jumped in after us. We hit the ocean with a splash, and I tried to keep the icy water from touching my skin.

"Just look at that sunset," I said.

Lon rolled his eyes.

Okay, so maybe Lon's not the friendliest person in the world. But he has other good qualities. I can't think of any at the moment, but they'll come to me.

The sailor turned on the trolling motor and we took off in the little boat, chopping along at a slow pace toward the ice. Before we made it very far, I noticed we weren't alone. A small metal raft moved out from among the icebergs and headed toward us. It was maybe five hundred feet away, so I couldn't quite make out the people on board.

Correction—the *ringlets* onboard.

The ringlets are new here. They used to live on the rings of Saturn. I guess they've been communicating with Earth since before I was born. Then two months ago, they suddenly decided to move here. They never said exactly why. Maybe it had something to do with all the radiation and space debris out there. Or maybe they just needed a change.

"Why'd they pick you two?" the sailor asked. "No offense. Just seems like a job for an ambassador or something."

Lon laughed. "Ringlets don't trust adults. Something about them being phony."

The sailor gave him a strange look. "Uh-huh."

I waited for Lon to tell him the rest of the story. How we entered the contest and were selected out of all the students in the country. He never missed a chance to brag, and this seemed like a perfect opportunity.

Instead, he made binoculars with his hands and watched the raft approach. After a few seconds, he turned his binoculars toward me. "Don't worry," he said. "I got this."

"What do you mean, *you got this*?"

"Just let me handle the diplomatic mumbo-jumbo. I'm good at that junk."

"Or we can, you know, *work together*. Like they told us at the embassy."

He pretended not to hear me. Typical.

The ringlets in the raft were getting closer. There were two of them. They looked exactly like the pictures I saw in training. Their bodies were tall and thin and completely blue. Their heads, rectangular and wrinkled. I didn't remember seeing orange short shorts and moccasins in the official photos, but maybe that was the latest fashion.

The sailor cut the engine and we drifted toward them. "Good luck," he coughed under his breath.

The other raft slowed as it approached, and soon we floated side by side. The ringlet on the left handed a rope to the sailor, who used it to lash both rafts together. He said something, but I couldn't understand a word. His language sounded like a mix of whale song and rushing water. It was beautiful—just wish I knew what it meant.

"I'll interpret," the other ringlet said. "Greetings, humans. On this most glorious night, with the stars burning wonderfully in the heavens, we graciously--"

"Nice to meet you," Lon said. "So how can we help?"

The ringlets stared at one another. I'm sure they weren't used to being interrupted like that. I mean, it made me feel pretty awkward. But I'm sure he didn't do it on purpose.

"The king is angry," the interpreter said.

"He is?"

"Yes. He is angry because your world destroys our homes. He asks why you banish us to the place where the sea swallows--"

"Whoa, wait. Did something bad happen?"

The ringlets looked at me in disbelief. I thought about saying something, but I kept my mouth shut.

"Yes, our homes have been--"

"Destroyed? I'm so sorry to hear that, guys. How can we help?"

That was my cue. Time to take over before things got any worse. I stepped between Lon and the ringlets. "Sorry for the interruptions," I said. "I'm not sure what happened, but I'd really like to help. Is there anything we can do?"

"The king would like a meeting. He wants to avoid a war."

Whoa. Hold up. A war? They didn't tell me it was this serious.

I gulped. "I'd be glad to meet with him and work it all out."

"Please come aboard."

I stepped over to their raft. Lon started to follow me, but one of the ringlets held out a hand and blocked him. "The king does not welcome interrupters."

Lon looked surprised. "What? Man, I was only trying to help."

The ringlets didn't change their minds. And they didn't invite the sailor either. Just me.

Lon leaned his head close to mine. "Guess it's up to you now. So, uh…no pressure."

Thanks, Lon. Just the encouragement I needed.

I sat between the ringlets and we moved back toward the ice. Not sure exactly how we moved, come to think of it. I didn't see any source of power. No engine, no sails, no oars. Just two blue guys and a piece of silvery metal that seemed to have a mind of its own.

"We forgot introductions," the interpreter said. "I'm Mimas. This is Janus."

"Gwen."

"Good to know you, Gwen."

“Likewise. Nice to meet you.”

The ringlets seemed friendly. I wondered if I could trust them. I wanted to, but it was too soon. I only just met them, after all. Now I was alone on their raft in the middle of the freezing ocean. I hoped they weren’t leading me into a trap.

I thought about the contest. The announcement from the president about how the ringlets wanted to resettle on Earth, and how they wouldn’t accept meetings with dignitaries or diplomats or normal ambassadors. They wanted someone younger.

All the principals got together and decided a nationwide essay-writing contest was the way to go. Describe why you would be the perfect choice to meet the aliens. Use complete sentences. Cite at least three external sources. 8 to 10 pages, typed.

Just my kind of contest.

Some kid out in Alabama took first place. He got to be the space ambassador and meet the ringlets in a capsule orbiting Earth and everything. He learned how they needed ice and coldness, and found them a place to live down here on the icebergs.

Somehow Lon and I tied for second. It was nice to get the framed certificate, but not nearly as nice as meeting aliens face-to-face.

And then the ringlets contacted our embassy. They had a problem.

Not trusting adults, they wouldn’t say exactly what it was. But the embassy worker wanted us to be prepared for anything. So she arranged for us to miss school and go to two weeks of special training.

I learned all about being a diplomat. I learned how the ringlets hold their utensils and how they greet strangers. The way their society works, the way they pass laws, all that.

And I spent hours studying ice. When it freezes, when it thaws. How it interacts with cold water. Hot water. The temperatures needed for hail versus sleet versus snow.

It was a lot of pressure. Kind of intimidating. The ringlets came from a world of ice. They were experts. How could I possibly use any of this to help them? But I did my best to pay attention anyway.

Now it was showtime. The ringlets had a problem, and it was up to me to figure it out.

***

The iceberg was even more impressive up close. It had to be several miles wide. We floated toward its right side. Over here the ice rose maybe twenty or thirty feet from the ocean. As my eyes moved along its bumpy surface, I watched it rise gradually until it formed a regular mountain of ice in the middle. It was majestic, especially with the purple twilight throwing shadows on the snow and reflections in the black water.

On the surface, I saw ringlets moving to and fro. I caught glimpses of tiny houses here and there. Igloos, almost.

We floated to the sheer ice wall, and then our raft rose straight into the air. I didn't have time to be surprised. Before I stopped to think about it, we were already safe and dry on the snowy ground above.

A blue man waited for us. He stood in front of a massive snow fortress. It had icicle guard towers and snowy ramparts and an awesome icy drawbridge.

"This is the king," Mimas said.

The king didn't look any different than the rest of them. No special jewelry or anything. The same neon orange short shorts. But he looked pretty upset. His face was scrunched up, and he spoke to Mimas in a loud, deep voice. He might have been yelling, but it's hard to say. Their language made it sound more like a lullaby.

"You don't need to hear all this," Mimas said, "but I'll summarize. He's saying how this would never happen back home. We never worried about plunging into the abyss there. He considers this a personal attack by your planet."

Wait. Plunging into the abyss? What does he mean? There's no abyss here.

"Mimas?"

"Yes?"

"Can I see what happened?"

"Right this way."

The three ringlets led me down a snowy path that led right to an edge. The king swept his hand over nothingness. I looked from the jagged ice cliff to the frigid ocean below.

"Yesterday this was a village. Twenty homes. Now, everything is lost. The king says you must be prepared for retaliation."

"What about the people who lived here? Are they…"

"They survived."

"Thank goodness."

I stared at the water for a few seconds. I listened to the wind and the creaking of the ice. I shivered and stamped my feet to keep them from going numb. Meanwhile the ringlets looked like they were enjoying a day at the beach.

"You like the cold, don't you?" I asked. "I mean, where you used to live, it was all ice."

Mimas said something to the king, and the king responded with a hand gesture and shrill whistle. "It's one thing to be cold; it's another to be miserable. That's why we came here."

Suddenly I understood the short shorts. "So you wanted someplace warmer."

"Oh yes," Mimas said. "It was far too cold back home. Of course, cold was all we ever knew. At least until we started learning about your world."

It was far too cold back home. I guess it's always cold there. So cold that the ice could never...wait a second!

"I just thought of something," I said. "I think I know what's going on here."

Mimas mumbled something to the king, and the king sang a quick solo.

"The king says this better be good."

I took a deep breath. "I'll try. It's called ice calving."

"Calving? I'm not familiar with that word."

"What about cows?"

“Cows I understand.”

“Okay. Calving is when a cow has a baby. This is similar. Well, not really. Not sure why they call it that. But I’m getting off-track.”

They stared at me expectantly.

“Sometimes the edge of an iceberg breaks off and drops into the ocean. Boom! Just like that.”

The king frowned and whispered something to Mimas.

“Why?”

I took a breath and hoped I remembered everything correctly. “Different reasons. Sometimes the ice gets too warm, and some of it melts and starts to crack. The cracks go down inside the iceberg and make it weak. Eventually a chunk snaps off.”

“So it’s not an act of war?”

I smiled. “It’s all natural. That’s how this whole iceberg was born, if you want to think of it that way. It fell off a glacier somewhere.”

The king shook his head and muttered something to Mimas.

“Your ambassador told us *this* was a glacier. Now you tell us it *came from* a glacier?”

I felt like Mr. Simmons, my old science teacher. Except I wasn’t wearing a turtleneck and didn’t have a Mount Fuji tattoo on my wrist.

“Close,” I said. “You live on an iceberg. Icebergs are pieces of ice that break off from glaciers or other icebergs. Glaciers are giant collections of snow and ice that move really slowly.

They form when the ice and snow collects faster than it melts or breaks off."

They whispered for a few minutes before Mimas looked at me again. "The king is grateful for the explanation," he said. "He says the war is off."

"That's wonderful."

"Quite an interesting world, Miss Gwen. We never came across anything like ice calving back home. I suppose it was much too cold. Anyway, the king would like to ask your advice on a new place to live."

I didn't know what to say to that, so I just smiled.

***

The king invited me to dinner in his ice fort. I'm sure it would have been amazing, but I thought of Lon and the sailor waiting for me, getting cold. I had to say no. We had a nice conversation about real estate in Antarctica, then said our goodbyes.

Janus and Mimas walked me back to the icy cliff and we boarded the metal raft. It carried us down to the water, and soon we pulled up beside the ship-to-shore boat. Lon and the sailor waited for me.

"I hope we meet again," Mimas said.

I waved at them. "Me too."

Their raft took them back toward the ice. A few seconds later, they were gone.

Lon looked cold and grumpy. I guess I couldn't blame him. He'd been waiting out here a long time. "So what happened?" he asked.

"Well, the ringlets had a problem. They thought we were attacking them, but once we got to talking--"

"Got it," he said. "Well, that was easy enough. Can't wait to tell Miss Martinez how we saved the planet and everything."

The sailor turned around and gave Lon a funny look. Lon lowered his head and mumbled into his mittens.

I sighed and wondered if I would ever see the ringlets again. Maybe they'd invite me to their new place in Antarctica. I could learn some of their language, explore the snowy hills, and do some serious sledding. Yeah, that would be sweet.

But right now, it was time to get back onboard. Before I did, I took one last look into the darkness. The giant iceberg eclipsed the yellow moon and turned the night sky into a spooky French painting.

It was pretty epic, I gotta say.

***

Justin Short lives in Kansas. His fiction has previously appeared in places like The NoSleep Podcast, Jerry Jazz Musician, and LOLcraft: A Compendium of Eldritch Humor. Visit him online at www.justin-short.com.

Art Credit: Breaking the Ice artwork by Erin Colson.

"Breaking the Ice" first appeared online in *Issues in Earth Science*, Issue 8, November 2017.

Chase Your Own Rabbits!

Question: Find something in the story that you wonder about, or want to challenge, or would like to explore more fully, and ask a scientifically-investigable question about it. As you consider things you wonder about, remember that not all questions can be investigated scientifically. For example, for the story *Breaking the Ice,* a question like "How fast does a glacier move?" might be more likely to stir a Google search than a field study of glacial movement like that once done by Louis Agassiz (as we discuss in *Learning to Read the Earth and Sky*, 2016, NSTA Press). "How can a glacier move, since ice is solid?" is an interesting question that might be difficult to address in the classroom with real ice, but might be addressed using an analog material that behaves like a solid most of the time but can ooze slowly over longer periods of time. Try to find a question that can be investigated scientifically with resources that you have, or can gather, in the classroom.

The question we, Russ and Mary, address in our Learn by Example for the story *Breaking the Ice* is the following:

"Is the claim in the story that higher temperatures make ice weaker really true?"

You can choose this question as a seed-question for your own investigation or come up with your own rabbit to chase! There are lots of possibilities in Breaking the Ice for questions related to ice, including questions about buoyancy (since they are on a floating iceberg), ice calving, glacial movement, or even climate change.

Feel free to write down your question here:

Vague Notion: As you prepare to plan your investigation, first get a vague notion of what you plan to do—including guidelines on whether it will

involve experimentation, data-interpretation, modeling, arguing from evidence or some other practice of science.

The vague notion that we started from for our Learn by Example for this story is the following:

Experimental design is something that we often get little practice at because we usually simply follow someone else's design. In this investigation, we plan to design an experiment, using common materials, to address the question of how temperature affects ice strength. Our goal will be to focus on designing and implementing the experiment as best we can with our materials and then evaluating whether or not our results are precise enough to give us an answer (by looking at our results, not by comparing with an expected textbook answer).

If you chose to go with our question for your own investigation, you can also choose our starting vague notion, or chase your own rabbit.

Feel free to write down your vague notion here:

Detailed Planning: Once you have your question and a vague notion in place, you can begin to fill in the nuances, complexities, and details of your plan. There is no recipe for your plan or a predetermined set of steps to follow, because your plan ***is*** the recipe or set of steps that you are creating for this particular investigation! Sketch out your investigation, anticipating as best you can what challenges you will encounter, and include as much detail as you are able (realizing that your investigative plan may not survive the first results of actually doing it.) Be sure to think about what background information you might need and how to get it, techniques you may need to research, the overall sequence of things you will need to do, and all other aspects of your investigation! You will likely need several pages to document your plan.

Follow a Question

Asking a question, choosing a vague notion of how to pursue that question, and then developing a detailed plan to pursue it is just the start of investigating. You actually have to do your investigation! During the course of doing your investigation, you will find out what works and what doesn't work. You may realize that you need more information, or different experimental tools, or even that your initial working model was so far off that your plan for investigation quickly falls apart. Authentic investigation involves iterative thinking during which your original experimental plan is revised, your initial interpretations updated, and your working model revisited. All of this is the whole point and practice of authentic investigation!

Keep careful notes on your experimental methods, interpretations, and models as well as your revisions as you work through your investigation.

Learn by Example (for Breaking the Ice)

Our Starting Question: "Is the claim in the story that higher temperatures make ice weaker really true?"

Our Starting Notion: Experimental design is something that we often get little practice at because we usually simply follow someone else's design. In this investigation, we plan to design an experiment, using common materials, to address the question of how temperature affects ice strength. Our goal will be to focus on designing and implementing the experiment as best we can with our materials and then evaluating whether or not our results are precise enough to give us an answer (by looking at our results, not by comparing with an expected textbook answer).

Introduction

The plot for *Breaking the Ice* turns on the idea that the warmer temperatures of Earth cause glacial ice to break more easily compared to ice on the Ringlets home world.

The Ringlets think that Earth is attacking them when their village falls into the sea with a calving chunk of ice, as implied in the following passage.

The king swept his hand over nothingness. I looked from the jagged ice cliff to the frigid ocean below.

"Yesterday this was a village. Twenty homes. Now, everything is lost. The king says you must be prepared for retaliation."

How can Gwen convince them that the calving of ice is a natural process on Earth, even if they are not used to it back on their home planet? She gets an idea for how to proceed when she reflects on what they said about the temperatures back home.

It was far too cold back home. I guess it's always cold there. So cold that the ice could never...wait a second!

"I just thought of something," I said. "I think I know what's going on here."

Warmer temperatures make the ice weaker, she explains. Is this true? Can we test it?

One way the ice can fail is to simply melt, but the calving of ice from a glacier involves more than simply melting. Does warmer temperature literally make ice "weaker" and more prone to break?

That is a question that can be addressed experimentally. Experimental design and implementation take a lot more work than discussion, calculations, or interpretation and modeling of existing data. That makes our investigation for *Breaking the Ice* one of the most ambitious of the Learning by Example investigations in this book.

A Short Primer on Experimentation in the Classroom

1) Designing an Experiment

The practice of designing an experiment, one aspect of an experimental investigation, is one of the most important practices of science. However, it is also one which students will have very little, if any, experience doing. Typically, we teachers perfect the experimental design on their behalf prior to assigning it to them.

When I (Mary) consider the experimental investigations my students did over my career, many of them were ones in which I gave students a pre-made experiment to follow like a recipe. I used these scripted labs sometimes for safety reasons, sometimes so students could finish a task in one class period, sometimes because I feared that student-designed experiments would be too "rough" to provide meaningful data. The purpose of the results from these experiments was typically to confirm, for ourselves, a key science idea we'd learned about during a lecture or discussion.

However, one consequence of this approach is that students don't ever learn to design experiments, or redesign them based on the results of their first efforts. They don't know how to constrain their variables, address experimental challenges, or even figure out whether or not their experimental results make sense. Because students don't design their own experiments, their analysis and interpretation of the data is skewed toward whether they got the "right" answer as determined by the answer key or textbook, and they don't engage with evaluating the nature of their experiment itself. This makes it difficult to encourage students to truly figure out what their data are telling them. The practice of designing an experiment engages students at an earlier stage in an investigation and sets them up to better know how to interpret their results.

Although experimental design is extraordinarily difficult to carry out in the classroom, learning how to invent a science experiment (and interpret the data) is at least as important as learning the theories—the "content"—of the discipline.

2) Interpreting Experimental Results

Data interpretation in the classroom is notoriously difficult because the results of any classroom experiment will never be exactly what we, or the textbooks, expect. I (Mary) remember trying to mentor students in interpreting their data, and struggling with how to both interpret the data myself and how to help my students interpret it. One key idea in moving toward authentic data interpretation is to get past the idea of evaluating whether or not an experiment "worked."

What do we mean by "the experiment worked?" Most often, we mean that the experiment yielded the result that we already expected, the result that the book, or the teacher, or the lab manual says we should get.

However, authentic science researchers don't know the results of their experiments ahead of time. Presuming that research results should be judged by how well they match the teacher's answer key undermines our ability to truly understand the nature of science: a knowledge-building enterprise that relies on empirical evidence, not answer keys.

Typically, in the college classroom, student-written reports include phrases like "the experiment failed due to human error." This failure to more thoroughly analyze the experimental results and consider what the results might mean, or exactly what went "wrong", reflects an answer-key driven experience with science in middle and high school. Real science is usually quite confusing and foggy at first, and there are no answer keys. The researcher needs to figure out what the experimental results mean from the results themselves, not from a teacher with an answer key in their back pocket. How can we engage students in similar experiences?

One place to start is to push students beyond "failed due to human error" to examine the actual results of their experiments—not what they think the results should have been. In every experiment, what happens is what happens. It certainly reveals something about the universe and the results are certainly consistent with the laws of nature. What <u>exactly</u> went "wrong" with the experiment? What do the actual results tell us? These questions are at the heart of the scientific enterprise.

3) Choosing a Question for Experimentation

In the activity for *Breaking the Ice*, we put these two problems in educational preparation (doing experimental design and evaluating whether or not it "worked") together to create an activity in which we design our own experiment and then evaluate what the results mean. We further consider whether or not the results are consistent enough (high enough precision) to yield real and meaningful understanding about the universe.

How can we design an experiment to test the idea that ice on Earth is more likely to break than the much colder ice at the rings of Saturn where the Ringlets came from, using commonly available materials?

4) Developing an Experimental Plan (and testing ideas to see if they work as we expect)

A creative activity like inventing an experiment is not something that can be done "on demand" in a classroom, like you might ask students to explain a theory or recall information. Creative activities take time to think about, to pose and test, and then to develop solutions to experimental challenges (perhaps another reason it is done so little in the classroom). Thus, you can't expect students to design and do an experiment in a single class period. Instead, you might start the investigation by having students spend a few minutes over several days talking about possible ways to develop the experiment. As possible approaches begin to gel, you can encourage students (or help them) to gather materials to test whether some of the ideas will work. Students can then revise ideas (again, perhaps using a few minutes over several class periods). After a couple of iterations like this, students can plan their experiment and carry it out using larger class-time blocks.

To design an experiment, students will need to work through a number of ideas, including the following.

What are the independent and dependent variables implicit in the experimental question? (In this case, temperature of ice and the force at which that ice will fail, or break.)

How can the independent variable be experimentally controlled (in this case, how can the temperature of the ice be changed and controlled) and how can the dependent variable be measured (in this case, how can the force at which the ice breaks be measured)?

How can other variables and factors be held constant? In this case, students need to identify other factors that might affect the strength of the ice, like thickness of the ice, its width, impurities and defects, etc, and consider how those can be held constant over several experiments.

Our Experimental Design (using commonly available materials)

Temperature control—the independent variable:
We, Russ and Mary, are fortunate to live in an area where temperature falls significantly below freezing, and so did not necessarily need to get access to a walk-in freezer with temperature control. However, we didn't want to have to wait on the weather, so we decided to instead design an "experimental chamber" in which temperature can be varied. We controlled the chamber temperature using mixtures of ice and salt in an external water bath and adjusted temperature by changing the proportion of salt. Adding salt to ice lowers the freezing temperature. (Side note: this chemical relationship allows us to melt road ice and to freeze ice cream in homemade ice cream freezers.)

Our experimental chamber is shown in the picture below, in which salt and ice are added to a Tupperware container inside a Styrofoam box. The Styrofoam provides some insulation from the warm room conditions. We placed a metal can inside the salt-ice mixture to provide a dry area in which experiments could be done. The metal is a good heat conductor and thus allows the cold conditions of the salt-ice mixture to lower the temperature inside the can. A Styrofoam top to the can limited mixing with warmer air in the room. The salt used for the example experimental design described here and below was a mixture of Ca chloride and Mg chloride purchased at Walmart (Prestone Driveway Heat). This product is used to melt driveway ice down to -32°C (meaning the equilibrium temperature for a mix of salt and ice is about minus 32°C). Take note that this is not edible table salt and safety considerations are listed on the container. Temperature was monitored inside the experimental chamber prior to each experiment using a digital thermometer as shown in the second picture below.

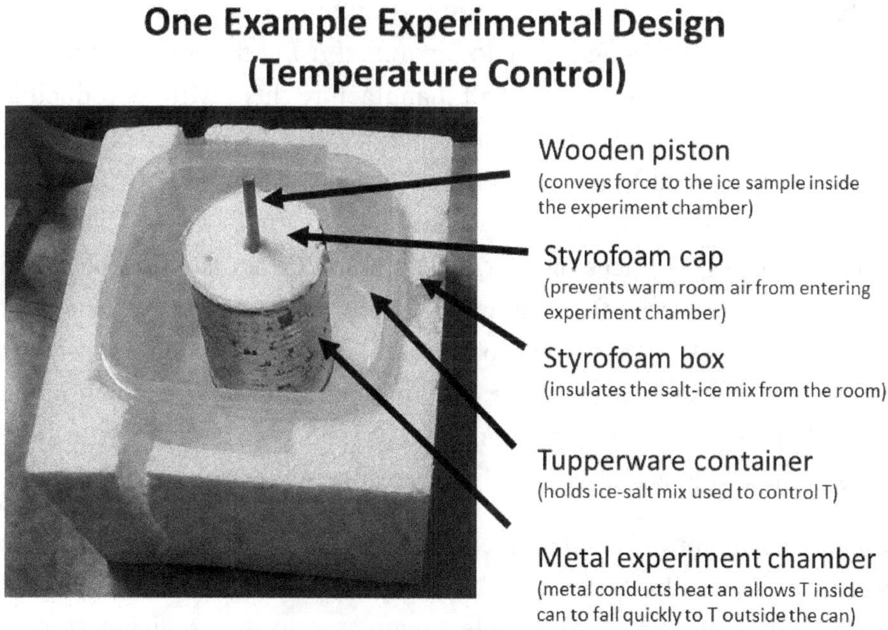

A thermometer was used to measure the temperature inside the experimental chamber prior to each experiment

Force measurement—the dependent variable:

We had to "try out" a few ideas before finding a solution to measuring force. Our first experimental vision involved using thin slices of ice (so thin that we discovered we couldn't manufacture them with reproducible thicknesses) with corresponding small amounts of force needed to break the ice. We thought that a small 16oz food scale should suffice. After some trial and error, we discovered that 16 oz of force was nowhere near enough to break the slices of ice we were able to manufacture. Thus, we gravitated to a mechanical bathroom scale allowing measurement of forces up to 300 pounds. The bathroom scale was placed against the wooden piston extending into the experimental chamber as shown in the picture below. Force could be monitored as it was gradually increased up to the point of ice breakage, and the force at breakage recorded.

Some mechanism was needed to convey the force to the ice chunk inside the experimental chamber. We chose a quarter-inch wooden dowel rod for the piston to convey the force (again, this was a revision from a previous idea using a lollipop stick which proved to not be strong enough).

One Example Experimental Design
(Force Measurement)

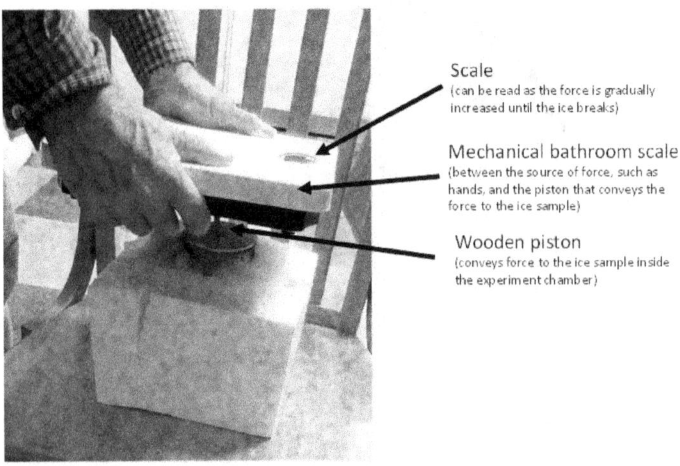

Control thickness and other dimensions of ice

As mentioned above, we initially thought we could produce thinner, weaker pieces of ice. Our first plan was to freeze ice in an ice cube tray and then cut thin slices, or perhaps slivers, with a hacksaw. However, the brittle ice fractured irreproducibly as we cut it with a hacksaw and this approach had to be abandoned.

Instead, we purchased an ice cube tray with a flat bottom (so ice thickness would remain uniform across the ice sheet) and only partially filled it with water, creating thin layers of ice of uniform width, length, and thickness. To keep thickness uniform, we measured 2 tsp water into each ice cube block (each with dimensions of 1.75" x 1.75", yielding, once frozen, ice sheets that were 0.2" in thickness).

Control the leverage that the force has on the ice.

The thickness of ice is not the only factor that influences its break-strength. The geometry of how the force is applied and the leverage on the ice also play a part. For example, imagine that you have a slab of ice laying across a 1" gap, and then the same thickness slab laying across a 12" gap. Clearly, the slab across the 12" gap will be easier to break.

To keep the gap constant, we laid the ice slab across a 1 1/8" wide circular hole drilled in a wooden block. The ice was centered over this hole, and the piston conveying the force to the ice was centered above the hole and the ice. The hole drilled part way through the wooden block at the bottom of the experimental chamber is seen in the picture below.

A hole drilled in a wooden block inside the Experimental Chamber provides a uniform gap for the ice to bridge.

Our Experimental Results and Interpretation

Once the experiments were completed, we needed to tabulate and/or graph the results to aid in interpretation. The results for the example experimental design reported here are given below. Note that "water-ice only" and "saltwater-ice" refer to the kind of bath in the Tupperware container used to maintain ice-slab temperature.

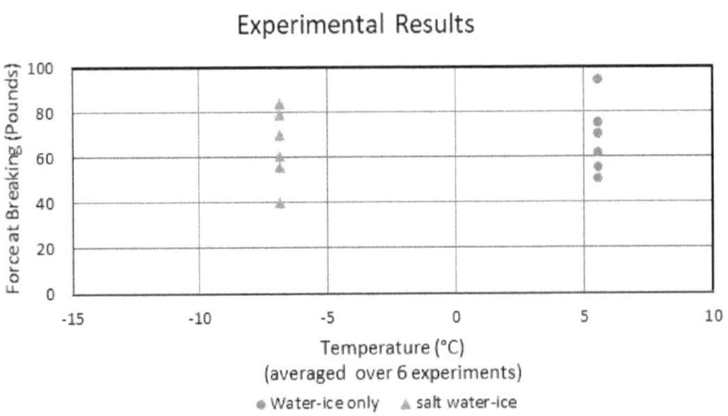

Specific questions that can be addressed with these results include the following.

1) Do the results indicate that the strength of the ice increases with decreasing temperature as implied by the story?

   In this case, no. The average force to break the colder ice (orange triangles) was actually 1.7 pounds *less* than the warmer ice (blue circles). What's more, as seen in the graph, the difference between the maximum and minimum force to break the ice, at either temperature, was more than 40 pounds! This means that there was more variation among trials at the same temperature than there was variation between the force required at the two different temperatures. Take note that this interpretive analysis required that trials be repeated to establish a baseline *uncertainty* in the measurements. Evaluation of experimental and measurement uncertainty (which can be inferred from the variation in a set of independent measurements) is a fundamental approach in all science. Since we can't compare our results to an answer key, we have to let the experiments themselves inform us about how precise they are. The range of variation, reflecting the uncertainty, can be seen graphically in the image above.

2) Do the results indicate that the strength of the ice decreases with decreasing temperature, the opposite implied in the story?

   Again, no. The average breaking force was indeed lower at the lower temperature (65 pounds versus 66.7 pounds). However, as can be seen from the graph, the variation in measurements was greater than this difference, meaning that the difference is not statistically significant. For those who like math statistics rather than graphical statistics, the standard deviation of the average at the lower temperature is 6.6 pounds and at the higher temperature is 6.5 pounds. A difference of 1.7 pounds of force at two different temperatures is thus significantly less than the standard deviation of the measurements at a single temperature.

3) So, can we say anything at all based on our experimental results?

   Of course. We can say that any change in strength with temperature, if it exists, has to lie within certain bounds. The slope of the change in strength with temperature must lie within the approximate bounding slopes shown in the graph below. We can also say that our experiments *did not* confirm that strength increases as temperature falls (although that might still be true within the limits shown in the graph below).

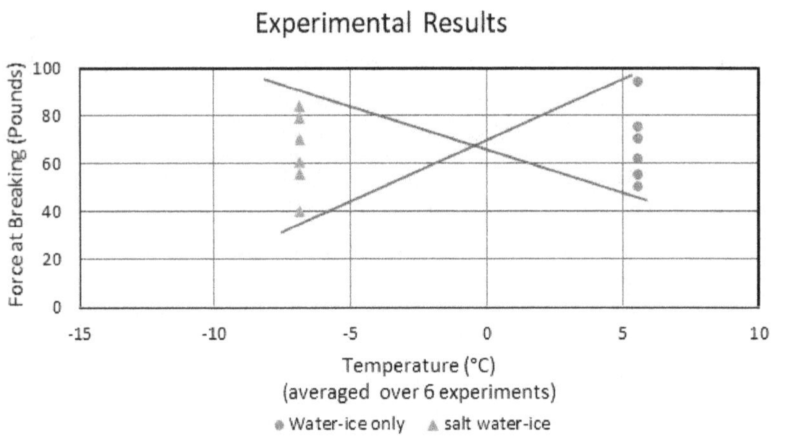

Summary of Questions our Experiment Addressed

Does the variation in our results indicate that our precision is high enough to distinguish a "real" effect of the independent variable (temperature) on the dependent variable (strength of the ice)?

Our answer was "No." However, if we had higher precision (meaning the data for each of the two temperatures had less variability), a next step might have been to consider whether, if there is a significant change in strength with temperature, does that effect indicate that strength increases with decreasing temperature or that it decreases with increasing temperature.

Had we concluded a change in strength with temperature, our next question might have been to ask whether our results made sense. Were there any other factors that might have caused the results to "seem" right but not be (for example, if the warmer ice block is melting during the course of the experiment, it would get thinner, and thus easier to break, which gives the "expected" result but for the wrong reasons).

Guidance for the Next Investigation

In real-world research, experimental progress is often slow, with researchers learning from previous work and moving forward with new approaches, thus making even "failed" experiments of great value. The next researchers to pursue this study can build on our results! As Isaac Newton said, we stand on the shoulders of giants—we build on the results of the people who came before us.

In the example experimental design reported above, improvements might include the following.

- Finding a way to increase the difference in temperature. A greater temperature difference increases the chance that variation in strength due to temperature can be detected above the uncertainty in the measurements. Better insulation, especially over the top, would make colder temperatures possible using the same salt-ice mix as used above, down to a theoretical limit of -32°C. The theoretical cold limit for ice and table salt is -21.1°C.
- More time might be allowed for the ice to equilibrate with the air temperature inside the experimental chamber (we only allowed 2-5 minutes, and in many cases the chamber temperature may not have reached equilibrium).
- Better control of where the piston strikes the ice. If the piston was not exactly centered above the 1 1/8" hole, and the ice centered over the hole, then the 'leverage' on the ice would be different, possibly accounting for some of the variation in measurements observed.

Comparing our Authentic Classroom Research with Published Research Science.

The shear strength of ice at various temperatures has been reported by Shun-ying Ji, Hong-liang Liu, Peng-fei Li, and Jie Su in the article "Experimental Studies on the Bohai Sea Ice Shear Strength", published in the *Journal of Cold Regions Engineering* in December 2013 (their Figure 4). Shear strength is the kind of strength that we measured where the force tended to "shear" the ice, rather than compressional strength, where the force is compressing the ice, or extensional strength where the force is pulling the ice apart. Their results are shown in the graph below.

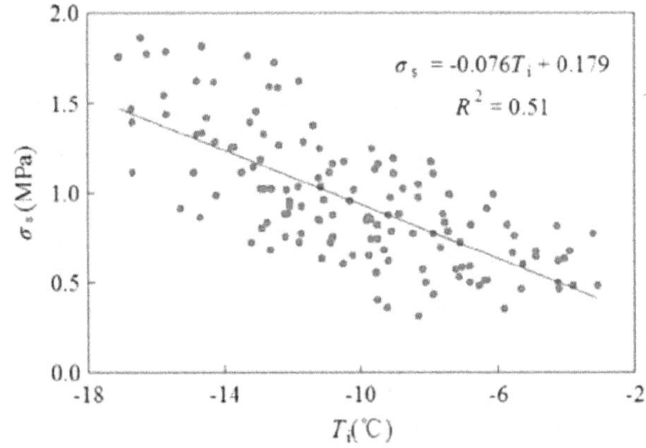

These researchers find a statistically significant effect of temperature on ice strength. However, notice that their results also *do not confirm* that lower temperature results in greater ice strength at the temperature range of our experiments above. The trend of their data is flat in this temperature region, and their results do not show a statistically significant increase in strength until the temperature is below -8°C. Notice also that their statistical uncertainty is as large (or larger) than our experiments done with a bathroom scale! For example, their measurements at -6°C range from less than 0.4 to 1 megapascals, a range of more than a factor of 2, similar to the variation from 40 to a little over 80 pounds in our experiments at -7°C. The researchers overcame their uncertainty by making many measurements over a wide range of temperatures, which allowed them to discern a statistically significant trend of greater strength of ice at lower temperatures. So, the story, *Breaking the Ice*, got it right after all!

Reflection on Learning

The goal of this book is to practice investigation itself, not to focus on the information and ideas you understood better through investigation. If traditional learning outcomes are not the goal, then what have you learned? Taking some time to consider what you gained from your investigation, reflecting on the practice of science *metacognitively*, can help you hold onto what you have gained. For example: What questions did you ask? What observations and interpretations did you make? Did you *do* modeling, rather than just learning or applying a model?

Real investigation is fluid, creative, and defies easy categorization. Even so, it might be helpful to consider the following categories of investigation offered by the Next Generation Science Standard's science and engineering practices (NGSS Lead States. 2013. Next Generation Science Standards: National Academies Press).

- Asking questions and defining problems
- Developing and using models
- Planning and carrying out investigations
- Analyzing and interpreting data
- Using mathematics and computational thinking
- Constructing explanations and designing solutions
- Engaging in argument from evidence
- Obtaining, evaluating and communicating information

Notice that we have not numbered these practices. They are not steps in a sequential process, but rather components of an integrated and iterative process. Also recognize that this list of practices is **a** way to organize and understand the investigative practice of science, not **the** way. As you go through this reflection, be flexible in thinking about what aspects of your own investigation might be practices of science, because these practices can show up in unexpected ways. Remember, don't think so much about

what you learned or understood, but rather think about what you initiated, did, practiced or experienced.

On a separate sheet of paper, jot down some things that have strengthened your ability to pursue investigative thinking. Give examples of how you engaged in particular, specific practices.

Exploring with you the experience that we, the editors, had in developing our Learn by Example section might provide some guidance as you think about your own practices of science. The focus of our Learn by Example for "Breaking the Ice" resides almost entirely in experimentation, one of the hardest aspects of science to practice in the classroom because of the time that it takes and the relatively uncontrolled directions it needs to go in order to resemble an authentic investigative experience. In the classroom, most experimentation consists of following the steps of an experiment that someone else has developed, tested, and mistake-proofed for you. However, this approach bypasses most of the practices of science. Our experimentation engaged us in the following practices.

Asking questions and defining problems was a major practice for us, the editors, throughout this experiment. We not only asked the initial question—does temperature affect ice strength and how—but at every step in the experiment we had to recognize and define a problem such as how to maintain the ice at a known temperature, or how to measure the force required to break it, and ask questions to guide us toward solutions. All of the decisions and revisions in this process constitute **planning and carrying out an investigation**.

As we developed and revised our experimental approach, we had to continually **model** what was happening in our experimental chamber and consider through that modeling how various changes to our experiment would affect the outcome and the suitability of the experiment. That modeling was guided by initial experimental results which had to be **analyzed and interpreted**, often resulting in changes to our model and our experimental design as we engaged in **finding explanations and designing solutions**.

Once our experiment produced usable results, we again had to **analyze and interpret** the data, considering, for example, its statistical significance, a type of **mathematical thinking**. We compared our results to work by previous researchers, requiring **obtaining and evaluating information**, and then made final comments on the meaning of our results, what it tells us about the natural world, and how it compares to the previous work, requiring the **construction of explanations**, **arguing from evidence**, and **communicating information**.

The development and implementation of an authentic experimental investigation that does not simply follow a preset script provides a robust engagement with all of the practices of science. While experimental investigations such as this take a lot of time, and, in most cases benefit from mentoring by colleagues, partners, or teachers with more experience in experimental development and design, they provide a fantastic engagement with the nature and practice of science. It is no wonder that this type of experience, albeit at a higher level, comprises the capstone achievement for advanced degrees in the sciences.

Permafrost

By Kate Kelly

Can six arctic explorers save humanity after war induces a chilling of world climate?

The cold bit into me. I forced myself on, step by painful step through the deepening drifts. Numb toes. Numb fingers. Beside me Mitzi stumbled and dropped to her knees and the sledge we were dragging slid into the back of my legs. I reached out with clumsy fur-wrapped hands and tried to pull her to her feet.

"I can't go on, Oscar," she said, her voice half lost as the wind whistled round us, driving the snow against my goggles, frosting my view. Nearby some jagged rocks jutted out of the all-enveloping shroud. The rest of the world was smothered.

I straightened and looked around at the grey mist and bitter whiteness. The snow was coming down heavier, swept by the wind into mountainous drifts.

Soon the day would darken into night.

"We'll camp here," I said. It was as good a place as any, the rocky outcrop affording some shelter at least. I left Mitzi kneeling in the snow and I turned towards the sledge and started to unpack the small two-man tent.

We had ditched the other tent when Anna died, covering her corpse with it and piling on rocks. There had been no signs that anything other than us moved in the desolate place, but I wasn't going to take the chance that there might still be wolves or bears. They weren't going to have her.

We were way behind schedule. Anna had held us up, Mitzi and I taking turns nursing her, hoping. And now the darkness was creeping in, the days shorter, the cold more severe.

And I missed her.

I glanced over at Mitzi as I struggled with my cold numbed hands. The way she was weakening now, soon there would be only me. I blinked the thought away.

It couldn't be far now. We had to succeed.

The world was depending on it.

Once the tent was pitched and we had taken shelter from the driving blizzard, I fired up the stove to heat water. We still had enough fuel – enough to get us there but maybe not enough to get us back. But if what the elders had told us was true – if this really did turn out to be the Source of Life, then somehow we had to succeed. The future hope for mankind depended on us.

In my dreams everything would change and this endless winter and encroaching ice would be gone. The world would return to greenery and birdsong; sunshine and warmth that was now no more than a distant memory – a half forgotten dream. In reality I knew that the world had changed. But mankind had survived the ice before, and would again.

I peeled off my gloves and inspected my fingers, fearful of frostbite, but it wasn't as bad as I had feared. I wriggled into my sleeping bag for extra warmth. The water was close to boiling but Mitzi sat, listless, staring at the side of the tent. It puckered and billowed as the wind sucked at it. I could hear the tarpaulin flapping against the sledge outside, but I didn't have the energy to go out and secure it. Let it flap.

I poured the water onto our dried rations and waited for them to soak it up. Then I dished them out into bowls and handed one to Mitzi.

"Chicken au Provence," I said.

She took it without speaking.

"When we get to the Source of Life we shall drink pink champagne and eat smoked salmon and caviar," I said, desperate to lighten the mood. "I reckon we're close now. Really close."

She spooned the mixture between blue lips and glanced up at me.

"We don't even know what it is, not really."

At least she was talking again, showing an interest. Maybe she would be okay after all.

"We know it will save us. We know it's the last hope for mankind, for our whole planet. We know that our ancestors foresaw a time such as this. The Source of Life is their insurance policy, so that mankind might survive." I said.

"Some old records salvaged from the chaos of war. They're not even complete."

"The archive is damaged, yes, but what is contains, the histories, the science, all of that is accurate. I have no doubt that we will find what we seek."

Mitzi sighed and continued spooning her rations into her mouth. I watched as she forced herself to swallow. Then I fished out my notebook to double check my navigation calculations. We had made up some of the lost time, but exhausted ourselves in the process.

There had been a time, a brief window of technology, when navigation had been easy, satellite navigation that could pinpoint your location to within a metre. But that world was gone. They blasted the surface of this planet to bits in that crazy suicide of a war. And now here I was, with Mitzi, trapped in a world of white and grey, with only the stars and the sun and a compass to map our way. Volunteers for a mission that might not succeed, to bring hope back to mankind.

There was a sad irony to the war – fueled by the troubles of a warming world, famine and drought. Oh yes, the war had changed all that – turning the world to a frozen waste instead. As the ice had expanded, civilization had crumbled. Now only the tropics were habitable, their climate changed beyond memory. Life clinging on while the crops that once grew at those latitudes failed.

“We’re close, Mitzi,” I said. “Really close.” I smiled at her even though the smile wasn’t returned. “We’ll get there tomorrow I reckon.”

Mitzi turned away from me, rolling herself up in furs and sleeping bag. I quenched the light and lay back, staring at the darkness and listening to the beat of the wind outside. I wondered what we would find – what the Source of Life really meant. The records stated that it had been put there for just such a time as this, so that mankind could be saved. But would it still be there? Could anything up here have survived both the war and the endless winter? Doubts filled my mind in the dark, and not for the first time.

We were selected at random from the many volunteers, trusted with the task of saving our planet, our species. Sent on a desperate mission to this lonely place at the top of a desolate world. The last hope of humanity.

We had started with two sledges and an expedition party of six all those months ago, when it finally dawned on our leaders that the ice wasn’t going to retreat; that the freezing of our planet had gone too far; that without action we were sure to perish. Just as those billions had died before, first from the effects of the warming, and then from an outright global war. We had set out, hearts filled with hope. A hope that eroded with time. Now there was only Mitzi and me, and a single sledge.

We had left the other sledge a long way back, back when Anna was still with me. Ah Anna. How my heart ached each time I thought of her. Ached for the future we should have had. But our love had always been marred by this changing world, from the day we met as aid workers, taking food to the starving refugees who had fled the desert that Europe had become. We were lucky, escaping the worst of the war, surviving the bombardment, only for the skies to darken and the world to freeze.

I owed it to Anna to finish what we had started.

Many times on our journey I had been sure we would fail, no more so than as I watched my colleagues die one by one. Slow deaths. Painful deaths. Our party had dwindled until now there were only the two of us.

But we had made it. We were nearly there.

I closed my eyes and listened to the wind. I tried to imagine Anna beside me, the touch of her hand, her smile.

Tomorrow.

But a day turned into a week.

Mitzi woke with a fever and for three days I sat by her side, keeping her warm, spooning melted snow between her burning lips, waiting for the fever to break. I feared she would die like the others and I would be left to do this alone, but the fever broke and then I waited, feeding her the last of our supplies as the wind piled snow against the side of our tent.

I waited as her strength returned. More time, more delay. And we were so close.

The wind had died down when we finally emerged to stand beneath dull dark skies. The snow had stopped falling but the wintery bite was bitter. It felt as if each breath was freezing in my throat.

I clutched my compass and pointed with a fur clad hand.

"This way."

We walked, deep snow, up to our knees. Each step an effort, the sledge a dead weight behind us. And the doubts crept in. What if it wasn't here? What if we missed it and walked straight past? I checked my compass again, then looked again at the hillside before me.

And there it was, jutting out from the bare rock face, pale and unwelcoming, but exactly where they had told us it would be.

Mitzi paused, head tipped to one side. "Is that it? It doesn't look like much."

I felt like laughing. It wasn't what I had expected either, perhaps something grander, more imposing. But now that I saw it, concrete and angular, I knew that this was what we sought.

"That's just the entrance. The Source of Life is inside. You'll see."

I unharnessed myself from the sledge and led the way, ploughing through thigh deep powder. I reached the door and paused. Mitzi joined me as I ran my mittened hands over the metal and concrete.

"Can you get it open?" she asked.

"Of course." I found a handle and pressed, listening to the satisfying clunk of the mechanism within. But the chain and padlock held firm.

"Bolt cutters?" said Mitzi. We had come well prepared – one piece of kit that I hadn't ditched, and I was glad of my decision. The chain didn't stand a chance.

Inside, a long corridor sloped down into the depths beneath the mountain. I pulled the torch from my pack and gave the handle a quick wind, then let the beam play over the walls and floor. The layer of dust was undisturbed. The place was intact.

We were the first.

“Lead on,” said Mitzi and her voice was breathy, excited.

The tunnel led deeper and deeper. My breath formed small clouds of fog in the freezing air. Mitzi kept herself pressed close beside me.

“What is it?” she whispered. “Do you know what’s in here? Do you think it will take us out of this nuclear winter?”

“I’ve no idea about that, the records were incomplete. But so far everything has been accurate, the location, and the purpose.” I whispered back, even though I knew there was no-one else to hear. My voice sounded echoey.

Mitzi was smiling. “They foresaw this coming, all those years ago. They knew this would happen.”

“They did, and we’re lucky as a result. I only hope humanity can learn from its mistakes.”

“We’ve come so close to obliteration.” Her eyes were shining. I didn’t answer. I was thinking of Anna, of how she should have been here. But this is what Anna would have wanted. She wanted me to live, to go on and rebuild. I was about to save the world

The tunnel ended up ahead.

At first I thought it just stopped, but as we drew closer I realised that it in fact bent off to one side. Another tunnel. We turned and followed. My torch started to fade and I paused to give it another wind.

This tunnel wasn't so long.

The reinforced door at the end led into some sort of airlock, one door hissing closed behind us before the next one opened. We stepped out into a vast room.

I stood and panned the torch beam around. It swept over row upon row of shelving, all stacked high with boxes, different colours, different shapes, grouped together by type. I could not see where the chamber ended.

"What is it?" said Mitzi.

"I don't know. It's some sort of vault. Let's have a look around."

"Maybe it's everything we need to survive, inside these boxes," she suggested. She studied the label of one and frowned.

"It will be everything we need to survive, that much is true," I said. "I guess we need to work out what to do with it all. There's too much here for us to take back."

Mitzi inspected more labels. I walked along the rows, along one, right to the end, and then back along another.

Mitzi was still inspecting labels. "They're all the same," she said and she turned to face me, eyes dark shadows and I felt a chill of foreboding.

"What's wrong, Mitzi?"

She shook her head in despair. "This is it. This is the Source of Life. Everything our species needs to recover from a global catastrophe. But they didn't predict the form that disaster would take. They thought there would still be a world to save. But all that is left is a wasteland of ice. A winter that will never end."

I shivered at her words. “I don’t understand, Mitzi. What do you mean?”

“I thought we’d find something to turn off the winter, to send the glaciers into retreat and warm the world so that it would once more be habitable – so that our species could survive.”

“But that’s never going to happen, Mitzi. You know that.” I took a step away from her and felt the wall of this strange chamber, hard against my back.

“But none of this can save us,” she said.

I was starting to shake. “I don’t understand.”

“I’ll show you.” She pulled down a box and ripped it open. Inside were sealed boxes. “Hold out your hands.”

I did as she asked. Trembling hands. My pulse was racing, my breathing shallow. She tipped the contents onto my palm. Hundreds and hundreds of tiny grains.

“Spring wheat from America and Europe, winter wheat from the northern Europe, Canada, and the American Midwest,” she said. “All the boxes are the same. It’s just that the seeds are different types. A vast archive of all the food crops of the old world. It’s The Source of Life. Only we can’t do anything with it. Nothing grows any more. What use are seeds to us?”

I let the grains fall from my hands onto the frozen ground. Mitzi stared at them in dismay.

"Winter wheat from Canada?" I whispered, and smiled. "Must be cold tolerant seeds."

***

Kate Kelly is a marine geologist by day and writer by night, writing fiction that draws on her interest in science. This story is set in a future scenario where a nuclear winter has tipped the planet into a new snowball earth.

Art Credit: Permafrost artwork by Erin Colson.

"Permafrost" first appeared online in Issues in Earth Science, Issue 11, August 2019.

Editor note: There is a real-world seed vault situated on the Norwegian island of Spitsbergen which, like the vault in the story, is a preservation of seeds made as insurance against global disaster.

Chase Your Own Rabbits!

Question: Find something in the story that you wonder about, or want to challenge, or would like to explore more fully, and ask a scientifically-investigable question about it. As you consider things you wonder about, remember that not all questions can be investigated scientifically. For example, for the story *Permafrost,* asking how mathematical modeling predicts climate change in the future gives us either a problem for the search AI (which is not investigative) or a problem that is so mindbogglingly complex that researchers are also struggling to solve it even with huge supercomputers. Asking "How would cold world temperatures affect snowfall in the coldest areas?" could lead to experimental investigations of the solubility of water vapor in air or database investigations of precipitation as a function of temperature worldwide, although the question might need to be focused further so as not to encourage simply Googling the answer. Try to find a question that can be investigated scientifically with resources that you have, or can gather, in the classroom.

The question we, Russ and Mary, address in our Learn by Example for the story *Permafrost* is the following:

"What are some of the factors that affect climate and how do changes in those factors affect climate?"

You can choose this question as a seed-question for your own investigation or come up with your own rabbit to chase!

Feel free to write down your question here:

Vague Notion: As you prepare to plan your investigation, first get a vague notion of what you plan to do—including guidelines on whether it will

involve experimentation, data-interpretation, modeling, arguing from evidence or some other practice of science.

The vague notion that we started from for our Learn by Example for this story is the following:

Brainstorm the wide sweep of factors that might affect world climate and apply observations and experiences from everyday life and previous understanding of scientific ideas to constrain and predict what changes might occur. Apply mathematical relationships to predicting, roughly, the magnitude of some changes.

If you chose to go with our question for your own investigation, you can also choose our starting vague notion, or chase your own rabbit.

Feel free to write down your vague notion here:

Detailed Planning: Once you have your question and a vague notion in place, you can begin to fill in the nuances, complexities, and details of your plan. There is no recipe for your plan or a predetermined set of steps to follow, because your plan ***is*** the recipe or set of steps that you are creating for this particular investigation! Sketch out your investigation, anticipating as best you can what challenges you will encounter, and include as much detail as you are able (realizing that your investigative plan may not survive the first results of actually doing it.) Be sure to think about what background information you might need and how to get it, techniques you may need to research, the overall sequence of things you will need to do, and all other aspects of your investigation! You will likely need several pages to document your plan.

Follow a Question

Asking a question, choosing a vague notion of how to pursue that question, and then developing a detailed plan to pursue it is just the start of investigating. You actually have to do your investigation! During the course of doing your investigation, you will find out what works and what doesn't work. You may realize that you need more information, or different experimental tools, or even that your initial working model was so far off that your plan for investigation quickly falls apart. Authentic investigation involves iterative thinking during which your original experimental plan is revised, your initial interpretations updated, and your working model revisited. All of this is the whole point and practice of authentic investigation!

Keep careful notes on your experimental methods, interpretations, and models as well as your revisions as you work through your investigation.

Learn by Example (for Permafrost)

Our Starting Question: "What are some of the factors that affect climate and how do changes in those factors affect climate?"

Our Starting Notion: Brainstorm the wide sweep of factors that might affect world climate and apply observations and experiences from everyday life and previous understanding of scientific ideas to constrain and predict what changes might occur. Apply mathematical relationships to predicting, roughly, the magnitude of some changes.

Introduction

The story in *Permafrost* implies a past period of global warming which caused economic and environmental stresses that led to war and the war resulted in a sharp decrease in global temperature and even more severe economic and environmental stress.

As we learn in the story:

There was a sad irony to the war – fueled by the troubles of a warming world, famine and drought. Oh yes, the war had changed all that – turning the world to a frozen waste instead. ...

.... The tunnel led deeper and deeper. My breath formed small clouds of fog in the freezing air. Mitzi kept herself pressed close beside me. "What is it?" she whispered. "Do you know what's in here? Do you think it will take us out of this nuclear winter?"

The science behind these changes in climate involve understanding how various factors affect world temperature. This understanding is based on studies of Earth's past climate as well as numerical modeling of the inflow and outflow of energy from the Earth. The inflow and outflow of energy from the Earth system are complex and at the frontier of modern research. Even so, we can engage with thinking about factors and feedback loops that affect energy flow. For example, we can think about how human activities in the past of the *Permafrost* world might have caused global warming and how war might have caused a subsequent descent into a global ice age. We can also consider other factors that affect climate.

In the activity below, the goal is to use our experience with the world around us to constrain how various factors might affect global temperatures, making this an exercise in applying observations from everyday life to a complex problem.

Challenge 1: A Game of What if!

The goal of this activity is to encourage us to ask questions about factors that might influence Earth's climate and then answer those questions through reasoning, observations we've made in the natural world, and/or online research.

The core idea in understanding changes in Earth's climate is that the energy that comes into the Earth system must also leave the Earth system, meaning that there is a balance of energy coming in and going out. When that balance gets out of whack, that is, when energy coming in and going out are not equal, then the Earth will heat up or cool down until the balance is restored. Even during times of climate change, the energy coming in and going out are very close to the same (otherwise we would see a much faster change in temperature). Although the factors that affect this balance are complex, they can be grouped into three simple categories, the understanding of which can help students predict the qualitative effect of various changes. Those categories are the following:

1) <u>Changes in the amount of energy that reaches the Earth.</u> For example, there might be changes in the Sun itself, or our solar system might pass near to another star, or there might be a nearby supernova which affects energy reaching the Earth.

2) <u>Changes in how much of the energy reaching Earth is reflected back into space without being absorbed.</u> Only the absorbed energy helps keep Earth warm. The amount of Earth's energy that is reflected is called

Earth's albedo. Increases in reflective materials like ice, clouds, or reflective particles in suspension in Earth's atmosphere can increase albedo and thus cool the Earth whereas a decrease in albedo, increasing absorption, will warm the Earth.

3) Changes in how quickly the energy that is absorbed by the Earth can reradiate back into space. All the energy that comes to the Earth eventually radiates back into space by black body radiation, the idea that all bodies emit energy at wavelengths and in amounts that depend on their temperature. Because of its temperature, Earth is "glowing" in infrared, a wavelength of light that our eyes cannot see. Greenhouse gases are gases that absorb infrared light, but not visible light, and thus delay the reradiated energy's journey back into space, keeping the energy in the Earth system longer and making the Earth warmer.

By considering the three categories above, we can predict the effect of a large number of factors on Earth's climate. For example, we might ask "what if the amount of ice at the poles decreases?" We can realize from our observational experience in the world that ice is reflective, meaning that decreasing the ice will decrease albedo and warm the planet, or we might search online using key words such as "what is the effect of ice cover on albedo?" Some *what if* questions might be more difficult to answer than others. For example, What if Earth's magnetic field switched polarity? How this might affect climate involves high energy solar radiation striking deeper into Earth's atmosphere and affecting temperatures at different levels of the atmosphere in complicated ways. However, some of our questions can be reasonably answered with our own reasoning and resources we can find online.

Let's make it a game! How many different factors can we identify that 1) have an effect on climate and 2) whose effect we can correctly predict?

What If Activity			
"What if" Factor Affecting Climate	Expected Effect	Type (1, 2, or 3)	Observations and Reasoning

Spoiler Alert:

***What If* Activity**			
"What if" Factor Affecting Climate	**Expected Effect**	**Type (1, 2, or 3)**	**Observations and Reasoning (other responses are possible)**
What if increasing temperature melts polar ice?	Greater Warming	2	I notice that snow reflects a lot of light, and days are colder with snow cover. This might decrease amount of sunlight absorbed at surface.
What if increasing temperature raises ocean temperatures?	Further Warming	3	Like the fizz goes out of a soda when it gets warm, CO_2 might go out of the ocean when they warm, increasing CO_2 in the atm and causing further warming.
What if increasing temperature allows plant growth farther north?	Decreased Warming	3	Plants use CO_2 to make woody material, removing CO_2 from the atm which might moderate warming.
What if increasing temperatures melts arctic permafrost?	Further Warming	3	Lots of carbon in permafrost might be exposed to oxygen in the air when melted, and oxidation of organics can create more CO_2 which would increase warming.
What if dust or aerosols from volcanic eruptions are injected into the atmosphere.	Decreased Warming or Cooling	1 or 2.	Reflective aerosols might increase reflection of sunlight before it ever gets to Earth's surface, decreasing temperature. However, some dust might absorb sunlight at different levels in the atmosphere causing a more complicated effect.
What if nuclear war injected dust or aerosols into the atmosphere?	Decreased Warming or Cooling	1 or 2.	See answer above.
What if black soot from diesel burning gets into the atmosphere or settles over the landscape?	Warming	3	Black diesel soot in the atmosphere looks very dark and so might absorb more sunlight, increasing warming, or soot settling on snow might melt it more, decreasing reflectivity and increasing warming.

What if warmer temperatures cause more water vapor to be in the air because more water vapor can exist in air at warmer temperatures?	Complicated	3	Since water vapor is a greenhouse gas and absorbed IR radiation, more water vapor might cause even more warming. However, if there is more water vapor available, then any cooling might cause even more condensation (like we get more condensation on our cold pop on a hot, muggy day) and so more cloud formation. Clouds are very reflective (they look bright when I fly over in an airplane) and so might decrease absorption.
What if people build solar panels on the top of buildings to capture incoming sunlight and convert it to electricity?	Complicated	2 and 3	Solar panels might be more reflective than black roofing tar and increase reflection, decreasing temperature. However, they also capture solar energy, keeping it present in the Earth system, which will eventually be converted to heat when the electricity is used. But the big effect might be to decrease the use of fossil fuels which produce CO_2, a greenhouse gas, that causes warming, and so would have a net cooling effect.
What if people plant more plants in urban areas and decrease the amount of concrete?	Complicated	2 and 3	Plants are very good at absorbing sunlight because that is how they get their energy. Thus, concrete is more reflective (more blinding on a bright day than grass for example!). Since the plants absorb more energy, that might increase temperature globally (although evaporation of water from the plant leaves tends to cool the local area and decrease the urban heat island effect). However, if the plants retain carbon in their woody material, this might decrease CO_2 in the atmosphere over time, at least until the plants die.

Challenge 2: A Game of How Much?

The goal of this activity is to give students a deeper *__quantitative__* *understanding of the energy balance that controls Earth's climate and to provide the opportunity to practice mathematical thinking.*

To some extent, we can model the effect of energy flow on global temperatures quantitatively, although student quantitative predictions are necessarily simplified and uncertain. Even climatic models by research scientists differ somewhat from each other and provide only our current best estimates of how various factors affect climate. This uncertainty is a function of the complexity of Earth's energy balance which in turn is a consequence of complex feedback loops that are difficult to predict from basic physics principles. Our second activity, a game of "How Much?" gives us the opportunity to think about how much a particular disturbance in the global energy balance might affect global temperatures by working through some simple numerical models based on two relationships.

Remember, the following puzzles and challenges are not intended to be systematic curricular steps through which students of science work toward defined learning outcomes. Rather, they are intended as aspirational examples of investigative thinking, appropriate at a variety of levels—some middle school, some high school, some for the expert teacher—to inspire investigative thoughts and increase understanding of what investigative thinking is all about. Find one or more that is right for you!

PUZZLE 1: Suppose that ice at the poles melted, causing the amount of light reflected to decrease to 29%. How much of a change in temperature do you expect?

Woah! How do I figure that?! The key is that most energy emitted back to space from the Earth is emitted by blackbody radiation, meaning the radiating energy depends primarily on the temperature (and not composition) of the body. Energy emitted by blackbody radiation is proportional to temperature to the fourth power (T^4). This means that the temperature is related to energy emitted by the expression

$$T \propto (\text{energy emitted})^{1/4}$$

Or

$$T_2/T_1 = (E_2/E_1)^{1/4}$$

If we increase the amount of energy emitted by 1%, such that $E_2/E_1 = 1.01/1.00 = 1.01$, then $T_2/T_1 = 1.01^{1/4} = 1.0025$). If it started out at $T_1 = 0°C$ (273K) then we might expect an increase to about 273K x 1.0025 = 273.6K, or about 0.6K (0.6°C increase or a little over 1 degree Fahrenheit).

The amount of energy that is emitted by the Earth is related to how much energy is absorbed because those two must remain in very close balance—that is, the amount of energy emitted by the Earth will be equal to the amount of energy absorbed by the Earth.

This gives us a way to figure out how a change in Earth's albedo (reflectivity) might affect global temperature (other factors ignored).

Right now, on average, about 31% of Earth's sunlight gets reflected, leaving 69% to be absorbed.

So, back to our question: Suppose that ice at the poles melted, causing the amount reflected to decrease to 29% (from the present value of about 31%). How much of a change in temperature do you expect? Hint: this will increase the amount of energy being absorbed by the Earth from about 69% to about 71%.

Spoiler Alert: Using the same reasoning and method as in the calculated example above, we can figure out how big of a temperature change to expect. Starting out with a world-average temperature of 288K (~15°C, a reasonable world average, instead of 273 Kelvin, ~0°C, as in the example above), gives us

$$T_2/288 = (71/69)^{1/4} = 1.0072$$

Or $T_2 = 290$

Roughly a 2 degree K increase (2°C, or 3.6°F).

PUZZLE 2: Suppose that dust and aerosols injected into Earth's stratosphere by a nuclear war causes Earth's albedo to increase to 37%. How much of a change in temperature do you expect?

Spoiler Alert: Again starting out with a world-average temperature of 288K (about 15°C, or 59°F), gives us

$T2/288 = (63/69)^{1/4} = 0.9775$

Or $T2 = 281.5$

Roughly a 6.5 degree K decrease (6.5 °C or 11.7 °F).

PUZZLE 3: Let's take a look now at the relationship between world temperature and CO_2 concentration in the atmosphere. If CO_2 in the atmosphere continues to increase exponentially as it has over the past 100 years or so, then we might expect the CO_2 to have doubled by 2100. How much change in temperature would we expect?

One expression that is used to model the effect of CO_2 on global temperature is the following, where ΔT is the change in global temperature at some future time t after time 0, ΔT_{doub} is the change in temperature that will result if the CO_2 concentration in the atmosphere doubles, $(CO_2)_0$ is the concentration of CO_2 at some time 0, and $CO_2(t)$ is the concentration of CO_2 at some future time t after time 0. The symbol "ln" is the natural log function. The x to the left of "ln" indicates multiplication.

$$\Delta T = \frac{\Delta T_{doub}}{\ln 2} \times \ln\left(\frac{CO_2(t)}{(CO_2)_0}\right)$$

So in "simple" terms, this expression predicts that temperature will increase as the natural log of concentration of CO_2 increases. The expression depends on the parameter ΔT_{doub}, an empirical value which is too complex to predict theoretically at the present time and so must be determined by observation. An estimate of this value can be determined by looking at how global temperature and CO_2 have changed in the geological past. Estimated values often lie between 1.5 and 4.5°C, based on past correlations between CO_2 concentrations and global temperature. However, in applying this empirical relationship, we should be aware of the difference between correlation and causation. Although we can see the correlation between temperature and CO_2 concentrations in high-quality data such as ice cores, these data alone cannot prove that the CO_2 increase is *causing* the increase in temperature. Thus, there is some uncertainty in how much variation in temperature can be expected based on a given change in CO_2—it might not be the same in the future as it has been in the past.

Given the relationship in the mathematical expression above, we can predict the effects of increasing CO_2 on global temperatures.

Now, let's get back to our puzzle: If CO_2 in the atmosphere continues to increase exponentially as it has over the past 100 years or so, then we might expect the CO_2 to have doubled by 2100. How much change in temperature would we expect?

Spoiler Alert: Choosing an intermediate value for our empirical value such that $\Delta T_{doub} = 3$, gives us the following relationship from the equation above.

$\Delta T = [3 \div \ln 2] \times \ln 2 = 3°C$.

PUZZLE 4: What if it doubled again in another hundred years?

Spoiler Alert: Again choosing an intermediate value for our empirical value such that $\Delta T_{doub} = 3$, we get the following relationship.

$\Delta T = [3 \div \ln 2] \times \ln 4 = 6°C$.

Combining changes in albedo with changes in greenhouse gases

Given the two mathematical models above, we can, if we choose, answer even more complex questions that combine the two mathematical relationships. For example, in the story *Permafrost*, how much of an increase in Earth's albedo is required to overcome the global warming effects due to greenhouse gases that was already ongoing? We would need to make a variety of estimates, such as: How much of a change in temperature would count as a global warming event? How much of a drop in temperature would be required to produce a global ice age?

Reflection on Learning

The goal of this book is to practice investigation itself, not to focus on the information and ideas you understood better through investigation. If traditional learning outcomes are not the goal, then what have you learned? Taking some time to consider what you gained from your investigation, reflecting on the practice of science *metacognitively*, can help you hold onto what you have gained. For example: What questions did you ask? What observations and interpretations did you make? Did you *do* modeling, rather than just learning or applying a model?

Real investigation is fluid, creative, and defies easy categorization. Even so, it might be helpful to consider the following categories of investigation offered by the Next Generation Science Standard's science and engineering practices (NGSS Lead States. 2013. Next Generation Science Standards: National Academies Press).

- Asking questions and defining problems
- Developing and using models
- Planning and carrying out investigations
- Analyzing and interpreting data
- Using mathematics and computational thinking
- Constructing explanations and designing solutions
- Engaging in argument from evidence
- Obtaining, evaluating and communicating information

Notice that we have not numbered these practices. They are not steps in a sequential process, but rather components of an integrated and iterative process. Also recognize that this list of practices is **a** way to organize and understand the investigative practice of science, not **the** way. As you go through this reflection, be flexible in thinking about what aspects of your own investigation might be practices of science, because these practices can show up in unexpected ways. Remember, don't think so much about

what you learned or understood, but rather think about what you initiated, did, practiced or experienced.

On a separate sheet of paper, jot down some things that have strengthened your ability to pursue investigative thinking. Give examples of how you engaged in particular, specific practices.

Exploring with you the experience that comes from this particular Learn by Example might provide some guidance as you think about your own practices of science. What practices of science might someone who works through these challenges experience?

The reader can find some traditional learning outcomes, for example, exploring the concept of albedo and how it affects climate, considering the concept of black body radiation and how it affects climate, and examining the effect of greenhouse gases such as CO_2 on blackbody radiation and climate. However, learning information or theories is not our goal. Our goal is to practice investigation. In what way does someone who works through the puzzles of this investigation gain experience with the practices of science? Here are a few examples.

Challenge 1 particularly focuses on **asking questions** about factors that might affect climate, **modeling** how those factors might impact climate, and then offering **explanations** or **argument from evidence** to support the modeling efforts. To pursue enough understanding to construct the models or offer explanations, one might need to **obtain and evaluate information** from reliable sources.

Challenge 2 focuses on **computational thinking** as it pertains to aspects of climate change, providing a mathematical foundation for the conceptual modeling of Challenge 1.

These exercises might engage with most of the practices of science, although, to be authentic, your investigation need not engage in all the practices.

Story by Scott Schad

Feedstock

By Scott Schad

When Frank Covalt discovers something in the mines of Mars that might threaten his employer, he has to choose between his job and his morality.

Dr. Frank Covalt sipped his quarter-liter of coffee and looked across the sea of solar collectors outside the windows of Margin Engineering.

It didn't look like a battlefield. He didn't feel like a soldier. But he'd been drafted by events out of his control and he'd chosen a side. What was company policy toward traitors? He was about to find out.

The company's offices hugged the Martian equator at 180 degrees longitude, surrounded by banks of solar cells. Each two-meter array patiently tracked the sun, cranking out 400 watt-hours of energy a day. They powered water-collectors methodically mining near-surface ice from the dusty Medusae Fossee rock formation. Frank did the math in his head, converting watt-hours into calories of heat, dividing that by the 80 calories it took to melt one gram of ice into water, and converting that into liters of meltwater per day. He came up with four. Each bank of solar cells could make four liters of water per day by melting the ice trapped in the Martian soil. It was the minimum amount of water that a human needed each day. If the solar cells harvested energy from the sun for half of each Martian day, then the coffee in his cup represented almost an hour of energy collection. It was expensive coffee.

Frank had been transferred here to solve the problem of how to make water cheaper. Then he had found a much bigger problem to solve.

His watch chimed. It was time to find out Margin Engineering's response to his discovery. Frank went to the closest stairwell that led down to the company's tunnel system. Offices, labs, and production facilities were arranged on the surface in concentric circles, connected by tunnels that formed the spokes of a big wheel. He entered the main tunnel and walked to Margin's administrative offices in the central hub.

Frank opened the Chief Administrator's door to be greeted by the icy stares of two men wearing corporate khaki. So this was Culbertson's idea of an 'informal talk?'

Chief Kellam Culbertson shifted behind his desk and nodded toward the doorway. “Here is Dr. Frank Covalt now, the man responsible for the report. Frank, these are corporate attorneys Hobbs and Dobbs.”

Frank suppressed a snicker and offered his hand to the guests. “Hobbs and Dobbs? No Knobs? Surely one of you is named Bob?”

“Remind me to pull this guy's psych profile before we leave today,” said khaki #1, Hobbs, who remained seated and did not extend his hand.

Khaki #2, Dobbs, was equally charming. “--and no JOBS either, at the rate you're going. What were you thinking, sending out a report like that?” He kept his arms crossed.

Frank gave them his best clueless expression and took a seat. It was always a pleasure to receive an impromptu visit from corporate. He suspected these guests had indulged in a big, expensed meal last night, followed by a special boost acceleration this morning during the bump-and-jump jug scooter from VMari. Many of the jug handlers were former scurry-rocket pilots, retired by the company just before their DNA profiles indicated a likely increase in insurance risk. Jug handlers therefore loved to receive corporate passengers, especially impatient, arrogant little snots like these two. He hoped they had loaded up on hydro shellfish last night. But there was no reason to be nasty to them, yet.

Frank chose to ignore #2 and to apologize to #1. “Sorry, couldn't resist.”

Culbertson frowned, “Resist.” Frank commenced resisting.

Khaki #2 fired again. “Well?”

Frank ignored him and looked straight at Culbertson. “Corporate attorneys, you say?” He received a solemn nod. “Did you call them?” Culbertson shook his head. Right. Frank decided that it was time to realign the conversation. “Who do I work for?”

Culbertson started, “Eh?”

Khaki #1 weighed in, “You work for us and you know it.”

Frank looked puzzled. “But I don't know anything of the sort. What is my pay grade and status?”

Culbertson's upper lip began to soften, sensing where Frank was headed. “You work for me and I work for Margin Engineering. I'll have to look up your rank and pay grade.”

Khaki #2 grabbed onto the tail end of the same thought. “We sign your pay vouchers!”

Frank inclined his head in a thoughtful nod. “Actually, you don't. Nor does Mr. Culbertson here. Margin pays a monthly stipend to Aerotek for my services, and THEY deposit credits into my account.”

“So...” Khaki #1 began.

“...and so I don't report to anyone in this room. Which means that you may take it up with Aerotek if my performance is unsatisfactory. You will also find that your contract with Aerotek is not binding upon me, as I am not one of their employees either. Geochemists are as rare as hen's teeth, in fact as rare as hens, Mars-side. I consult to Aerotek on my own terms and they are happy to have me.”

Hobbs and Dobbs shared a sneer.

Culbertson smiled painfully. “Perhaps,” he said, “Dr. Covalt could simply explain the reasons for his poor assessment of our Paradox project.”

“Gladly. May I proceed gentlemen?” Frank nodded toward the khaki twins. Hobbs flicked one hand in a “whatever” gesture.

“Very well. It has been known for some time that many regions on Mars contain subsurface ice. Mining this ice to turn it into water is an energy-intensive process. Indeed, our present mining industries were energy-constrained until I discovered deep reserves of liquid water. If these water reserves are intelligently developed, Margin can look forward to many decades of lucrative work.

"Aerotek renewed my contract in order for me to assay possible liquid water deposits in your stage three mining concession, or as you call it, your Paradox project. I chose to begin by comparing known reservoirs to establish a prudent baseline case.”

Hobbs, proving to be the more vocal of the corporate bookends, interrupted, “And so, after renewing your assignment, you began by wasting Margin's money, telling us what we already knew!”

“Not at all,” Frank remained unruffled. “A base case provides essential calibration for any forecast. It also clearly establishes knowns, which in your case, were not so.”

Culbertson coughed. “Margin was quite on top of water quality issues, Frank.”

“Yes sir, but only in a qualitative sense. Margin owns exactly one fluorescence spectrometer, which hardly provides a robust method of analysis. By expanding the scope of my examination I provided the means for you to measure the scale of what now looks like a planet-wide problem.” Frank paused while Dobbs looked at the ceiling and exhaled loudly, then he continued.

"I collected certain facts and mapped them. Without exception I found that in each region where Margin has discovered liquid water a distinct salinity profile exists. Whereas your shallow exploratory wells tapped fresh waters, your later and deeper wells found progressively higher salinities. Your own corrosion maintenance reports bear this out."

Culbertson shifted his overfed frame, creaking his chair. "Frank, we all know the history of water on Mars. There is no need to give us a miner's introductory course."

"Yes, we all know that the water essential for any expansion of our operations on Mars originally came from glaciation. Our best estimates put last-stage glaciation from 50,000 to 100,000 years ago. But gentlemen, glaciers consist of frozen fresh water. Where did the salt come from?"

"Oh come now," Culbertson frowned, "surely no one disputes that salt came from the bedrock itself as water flowed through it? Do you challenge that?"

"In your stage one mining area my work clearly shows that, in fact, salts dissolved out of the host rock. But in all other cases, and in the rest of Margin's concessions, no suitable rock formation exists to provide the salts. Further, it is only in the remaining cases that we find clathrates."

"What?" Dobbs shook off his bored stupor at the unfamiliar word.

"Gas hydrates. Think of clathrates as ice cages. Microscopically, they resemble linked geodesic spheres, each just big enough to contain one molecule of methane gas. We find them at the extreme updip ends of the water reservoirs, closest to the cold surface of the planet."

Culbertson addressed an aside to Dobbs: “We know all about these gas-ice hydrate deposits. They've never proven to be a problem for our operations.”

“And yet,” Frank said, “no one before me attempted to analyze the gas that they contain. Isotopically, the methane gas trapped in these small ice cages is identical to Archaebacteria-sourced gas on earth.”

“Who cares?” Hobbs was back on point.

“We should all care, Mr. Hobbs, because all lines of evidence converge. Take the organic carbon values of the seal rocks encasing your aquifers, for example. Some measurements exceed 20% carbon by weight. Or look at the obvious link between bacterial generation of methane and bicarbonate production. This ties to the increased alkalinity due to carbon dioxide being turned into bicarbonate at the bottom of your reservoirs.”

“Produced by aliens?” Hobbs sneered.

“Yes sir. By alien life forms comparable to Archaebacteria, living in subsurface communities in fresh water halos above the alkaline waters. I've spent the last two years testing water samples across your concession areas and they clearly show the alien biota as assemblages of flimsy cells and filaments no different from their counterparts on earth. The microbes appear to be eating the organic matter in the high-carbon shales which seal your aquifers. They excrete both carbon dioxide gas, which drives the bicarbonate formation, and methane gas, which ascends to take up residence in the ice cages.”

Hobbs now spoke calmly, trying to convince a child that the nicknack in his hands was too valuable to drop. “Do you realize the public relations nightmare that you have created? Despite our best attempts to confine your report to internal channels, it somehow reached our biggest investor group two days ago." Frank perfectly suppressed a smile. He knew a thing or two about unofficial channels.

Hobbs continued, "Margin's CEO was completely in the dark when they contacted him! He was sitting at his breakfast table when it suddenly lit up with dozens of inquiries about the establishment of "first contact" and how it would affect future operations. I can tell you that a small supernova went off shortly after that in the office of our lead counsel. We," he motioned to Dobbs, "are just the advance guard. And we have orders to bodily deposit you into the next orbital loop taxi to earth, minus your planet creds and travel visa."

Frank stiffened in his seat. He had expected fallout, but the speed and ham-fistedness of Margin's damage control surprised him. "Would you gentlemen excuse me for a second?" Frank tapped his watch to turn off a vibrating alarm. "I have to pick up my son now. May we continue in half an hour?"

Hobbs rose to his feet. Dobbs followed, reluctant, but deputized. Perhaps they intended to shadow him until he was safely put off the planet, but Culbertson reasserted control of the situation. "Your son has already been picked up, Frank."

Dr. Covalt remained composed. "Oh? By whom?"

"My secretary, Janet, picked up your son Angus as this meeting started. I didn't want us to be interrupted." Culbertson then smiled, in a new way that Frank did not recognize. Hobbs and Dobbs, also surprised, faced Culbertson with new respect at his demonstration of corporate loyalty. Culbertson's smile vanished as they resumed their seats.

Frank marveled at the swift stupidity of Margin's response. Culbertson explained that Aerotek's contract had been canceled. Unable to trace the route by which Dr. Covalt's report had reached the public, all contracting services were hereby terminated. Non-Margin employees were to be put off the premises and denied access to all mining concession sites. The deadline to vacate was the familiar EOBD, end of business day.

Culbertson paused to read the corner of his desk, then he elaborated Margin's position: "Frank, you may or may not have forced our hand, yet we have no choice but to force yours. You are here at the convenience of the company and you will leave at that same convenience. Our jug pilot will hand you over to corporate security, who will escort you to your quarters in vMari. From there through the enclave you are at liberty, but to save you time and worry your son will meet you at the taxi shuttle port. We want you off-planet by midnight."

Frank nodded to Culbertson. "You have my report. My work here is done."

He turned to address the two corporate hatchet men. "Your work, however, is just beginning. Your employer has had the honor of making first contact with an alien species. Apparently they do not view it as an honor. It's time to ask yourself, who do YOU work for? A company that plans to exterminate a helpless life form and lie to the whole earth about it? Aren't lawyers required to raise their right hands and swear fidelity to the truth when they pass the bar? Ask yourself how will posterity deal with the people who enable this unique genocide? Or, shorter-term, how will Margin Engineering deal with him," Frank pointed at Culbertson," or with you two, given what you now know? It's time to pick a side, gentlemen."

No one said anything for a minute. Dobbs finally broke the silence. “Corporate asked us to express their appreciation for the rigor of your work here, Dr. Covalt, and to tell you to expect a bonus payment at the termination of your contract. Our ex-pat placement group has also volunteered to help you locate a new position on earth, perhaps in a less charged environment, where you may continue your career without distraction.”

“I thank you gentlemen for your kind offer, but several universities previously extended open invitations to me, should I ever desire to return to earth. If they no longer represent viable options when I return, then I will gladly accept your help.” Frank rose with perfect dignity. Culbertson wore the smile-frown trained into all managers, and Hobbs and Dobbs puffed in relief. The matter had obviously resolved to everyone's satisfaction.

Corporate security accompanied Dr. Covalt on the long, bumpy ride to vMari. A bored security technician sat at the rear of the cramped compartment, keeping an overt eye on Frank as he sat near the jug-wrangler and his copilot. The security tech even followed Frank to his apartment, so Frank drafted him into helping pack his son's things.

He spent a few minutes parsing his meager store of possessions, chose a couple of traveling suits, then rummaged for his personal data clips until he remembered the technician saying that they had already been transferred to the shuttle. Sure. He fought a moment's impulse to anger, but the technician in the other room had not been the one to light-finger his personal files. It would do little good to raise the issue, since Mars was essentially one big corporate town. An appeal to higher, non-autocratic authority lay two years and a planet away.

The security tech called from the next room, “Hey, Dr. Covalt, what do you want me to do with this?” Frank looked through the door and saw that the man was holding a toy hydrofish tank. Little plastic bivalves clapped on the bottom of the tank, opening and closing rhythmically.

“My son is pretty attached to that. Do you think it would survive the shuttle ride up to the taxi?”

The tech looked it over. “These toys are pretty cheap. They sell them over at the visitor's post, but,” he held it up to the light, “I guess I could find a shoulder capsule that would hold it. You think it would leak?”

Frank smiled, “Maybe. But if it does make a mess then I'll be able to convince my son to throw it away. You know how kids are, how they get attached to things.”

The tech nodded. “Gee, I'm sorry that things turned out this way, Dr. Covalt.”

“It's for the best,” Frank said. “My late wife wanted my son to see earth someday.”

The technician, not knowing what to say, resumed packing.

A day later, swinging into the perigee of the taxi's looping transit to earth, Frank sat with his son in the small compartment they would call home for the next two years. “Now you will remember to feed them the special food?”

“Yes sir, I promise.”

“Good.” Frank patted his son's knee while they both stared at the little tank of Martian water, with plastic clams opening and closing on the bottom. "They'll be the last of their kind you know.”

“I know,” his son said. “It’s kind of sad.”

“And even though this is a pretty strong little house that we built for them inside your toy, they are still fragile. We'll raise them together, you and I, and make sure that they survive.”

“Yes Daddy. Are you sure they are in there?” The little boy peeked into the clear plastic tank.

“Yes, I'm sure.”

***

Scott Schad is a professional geologist and writer in Oklahoma. He is the author of the "Meteor Men" adventure book series for boys, published by Raconteur Press.

Art Credit: Feedstock artwork by Erin Colson.

“Feedstock” first appeared online in Issues in Earth Science, Issue 19, May 2025.

Chase Your Own Rabbits!

Question: Find something in the story that you wonder about, or want to challenge, or would like to explore more fully, and ask a scientifically-investigable question about it. As you consider things you wonder about, remember that not all questions can be investigated scientifically. For example, for the story *Feedstock,* asking the question "is Dr. Cavalt's calculation of the energy cost of his cup of coffee correct" might involve some online searching—not investigative, but could still be made investigative if the setting up of the problem and calculations are done yourself. Asking "why would Dr. Cavalt be looking for liquid water deposits rather than water tied up chemically in the rock and sediment" is certainly in interesting question, but it might end up becoming a discussion of economic cost and benefit or a Googling exercise in finding out the distribution or rarity of liquid water on Mars. Try to find a question that can be investigated scientifically with resources that you have, or can gather, in the classroom.

The question we, Russ and Mary, address in our Learn by Example for the story *Feedstock* is the following:

"What is the morality of mining, the evidence for life, and the interpretation of observations cited in the story?"

This is an extremely broad starting question, and we will pursue the question less substantially than most other questions in our Learn by Example sections. You can still choose this question as a seed-question for your own investigation, perhaps developing it further than what we do, or come up with your own rabbit to chase!

Feel free to write down your question here:

Vague Notion: As you prepare to plan your investigation, first get a vague notion of what you plan to do—including guidelines on whether it will involve experimentation, data-interpretation, modeling, arguing from evidence or some other practice of science.

The vague notion that we started from for our Learn by Example for this story is the following:

We have chosen to spread ourselves a bit thin for this story, initiating three limited investigations developed to a lesser degree than our previous projects—we might call them our plan *for investigation. Perhaps you can take one and expand it to a more complete investigation! The three investigative efforts consist of the following: 1) a discussion of the moral consideration of mining operations, on Mars or Earth, 2) an incomplete and untested experimental investigation of evidence for life, and 3) a consideration of the relationship between observational evidence and conclusions.*

If you chose to go with our question for your own investigation, you can also choose our starting vague notion, or chase your own rabbit.

Feel free to write down your vague notion here:

Detailed Planning: Once you have your question and a vague notion in place, you can begin to fill in the nuances, complexities, and details of your plan. There is no recipe for your plan or a predetermined set of steps to follow, because your plan ***is*** the recipe or set of steps that you are creating for this particular investigation! Sketch out your investigation, anticipating as best you can what challenges you will encounter, and include as much detail as you are able (realizing that your investigative plan may not survive the first results of actually doing it.) Be sure to think about what background information you might need and how to get it, techniques you may need to research, the overall sequence of things you will need to do, and all other aspects of your investigation! <u>You will likely need several pages to document your plan.</u>

Follow a Question

Asking a question, choosing a vague notion of how to pursue that question, and then developing a detailed plan to pursue it is just the start of investigating. <u>You actually have to do your investigation!</u> During the course of doing your investigation, you will find out what works and what doesn't work. You may realize that you need more information, or different experimental tools, or even that your initial working model was so far off that your plan for investigation quickly falls apart. Authentic investigation involves iterative thinking during which your original experimental plan is revised, your initial interpretations updated, and your working model revisited. All of this is the whole point and practice of authentic investigation!

Keep careful notes on your experimental methods, interpretations, and models as well as your revisions as you work through your investigation.

Learn by Example (for Feedstock)

Our Starting Question: "What is the morality of mining, the evidence for life, and the interpretation of observations cited in the story?"

Our Starting Notion: We have chosen to spread ourselves a bit thin for this story, initiating three limited investigations developed to a lesser degree than our previous projects—we might call them our plan *for investigation. Perhaps you can take one and expand it to a more complete investigation! The three investigative efforts consist of the following: 1) a discussion of the moral consideration of mining operations, on Mars or Earth, 2) an incomplete and untested experimental investigation of evidence for life, and 3) a consideration of the relationship between observational evidence and conclusions.*

Introduction

Life on Mars! We've been looking for it for nearly two centuries, and now we find it!

.... *"And yet," Frank said, "no one before me attempted to analyze the gas that they [clathrates] contain. Isotopically, the methane gas trapped in these small ice cages is identical to Archaebacteria-sourced gas on earth."* ....

"Produced by aliens?" Hobbs sneered.

"Yes sir. By alien life forms comparable to Archaebacteria, living in subsurface communities in fresh water halos above the alkaline waters.

However, not everyone in the story has the same reaction to the discovery of life, some wanting to suppress it, some not believing it, and some unsure how to go about saving that life from corporate self-interest. Feedstock addresses an issue of clear importance in the real-life modern world: the impact of human activities, mining in this case, on other organisms and ecosystems—a perfect opportunity for a classroom discussion. The story itself offers an oblique examination of evidence for life on another world—a perfect opportunity to consider experimental examination of the presence of life and what counts as evidence, conclusion, or simply biased presumptions in the story itself.

For this last story, we're going to present our *plan* for an investigation, rather than the results of that plan as we've done for the other stories. We break our plan into three parts: 1) Discussion of Mining Morality, 2) Experimental Investigation of Biological Activity, and 3) Identifying the evidence for Martian life in the story *Feedstock*.

Remember, the following puzzles and challenges are not intended to be systematic curricular steps through which students of science work toward defined learning outcomes. Rather, they are intended as aspirational examples of investigative thinking, appropriate at a variety of levels—some middle school, some high school, some for the expert teacher—to inspire investigative thoughts and increase understanding of what investigative thinking is all about. Find one or more that is right for you!

Challenge 1: Discussion of Mining Morality

The story, *Feedstock*, centers around the discovery of life on Mars that might be threatened by the mining operations ongoing there and the response of the mining company to that possibility. There are real-world current things happening on Earth that are very much like this scenario.

To set the stage, science, exploration, and technological developments always have consequences, both negative and positive. For example, the scientists who studied nuclear processes through the first half of the 20th century are often criticized because of the use of that knowledge in making nuclear bombs dropped on Japan during World War II. However, if no new understanding had ever been pursued because it carried potential negative consequences, then humanity would never have started the first fire, press-flaked the first flint knife, or sealed the keel of the first water vessel. All of these things have been used for negative purposes.

The terrible truth is that all discovery has been used for both good and bad. Should we therefore do nothing and discover nothing so as to avoid the bad? Or do we do it all so as to pursue the good? Different people may come down on different sides of this question, however, history has shown that those who pursue the discoveries will, in the end, likely win the day, whether good or not. Those who had fire, knives, and ships conquered those who did not (ouch, people should be nicer to each other, don't you think?).

The (somewhat similar) moral question for us today, that bears not only on the story *Feedstock* but on many modern challenges, is how much weight do we give our own human needs and wants, and how much weight do we give the existence of other creatures that may be affected by our actions?

Mining the floor of the deep ocean for manganese nodules and other resources is an issue not unlike the problem in *Feedstock*. Our hunger for ever-more technological devices and infrastructure drives a need for copper and other natural resources and that need turns our eyes to the oceans. Ocean areas comprise more than 70% of the Earth's surface and the deposits of the sea floor are a treasure trove of mineral resources. They are also a treasure trove of highly fragile and rare ecosystems. The extreme long-term stability of these deep ocean environments means that the ecosystems present there are not adapted for disruption, and the disruption of mining might destroy them. Is the value of the resources we might extract worth that loss?

Another issue similar to the problem in *Feedstock* is the potential for mining copper in our own state of Minnesota. Although home to one of the largest copper deposits on Earth, Minnesota has been known for iron mining. Iron mining is a relatively "environmentally clean" type of mining compared to the potentially-more-damaging copper mining that involves sulfur minerals that create acidity that might leach pollutants into the surrounding land and water. Native American communities, including the Ojibwe, are often particularly concerned about the potential impact of copper mining on wild rice, fishing, and other parts of the natural ecosystem of particular importance in Ojibwe culture. In Minnesota, one of our state earth science teaching standards includes integrating Native American views, such as Ojibwe, into the curriculum, and the discussion of the moral quandary in *Feedstock* might provide an opportunity to do that.

A third direction in which a discussion might go is toward the philosophical and moral aspects of economics. Capitalism, for example, is tied closely to the idea of encouraging people to act in their own interest, thereby maximizing productivity and innovation. However, selfish interest has to be balanced by group interest in some way, at the very least limiting the ability of one person to take from another in order to enhance their own prospects. Economists have grappled for centuries with the problem that selfish interest can sometimes deplete or destroy mutually-shared resources, such as water, air, and ecosystems, an idea that has come to be called the tragedy of the commons. Even Adam Smith, father of capitalist thought, recognized that a resource in which no particular owner has a vested long-term interest can be lost to society as a whole. In this context, ideas of morality, altruism, and unselfish action, often relegated to the humanities, become important in science and economics.

Preparation for discussion:

Do some searching online to learn about mining the sea floor and its ecological consequences. Get perspectives both from the point of view of mining companies wanting to capitalize on valuable resources that support a prosperous world and environmentalists concerned about the morality and long-term consequences of destroying ecosystems.

Search online for the positive and negative potential of copper mining in Minnesota, and also search for Native American perspectives on copper mining in Minnesota.

Search online for thoughts and consideration of issues of morality in economics, selfish interest versus group interest, the economic or moral cost to the use of resources at the expense of other species or ecosystem health, or the tragedy of the commons in capitalist philosophy.

Challenge 2: Experimental Investigation of Biological Activity

Our exploration of Mars has always included an interest in finding life there. When astronomers turned their telescopes on Mars in the late 1800s and sketched what looked like channels, some speculated that these were canals, constructions of an advanced society. Viking I, the first ship from Earth to successfully land on Mars in 1976, carried equipment to test the soil for presence of microscopic life. The last two decades of exploration on Mars has carried the explicit goal of looking for water as a prerequisite for life (as we know it), as though looking for rocks isn't good enough (our geologist-hearts go *grrrr*—we long to see the news headline "Rocks discovered on Mars!")

The Viking lander did a number of tests looking for metabolic activity, including looking for generation of gases that result from respiration, such as carbon dioxide (CO_2).

We might attempt a much simpler experiment in the classroom to try to detect biological activity in soil or sediment samples, perhaps trying out bromothymol blue as a CO_2 indicator.

Skeletal plan for a first experimental effort. Collect two cup-sized soil samples from a moist, organic rich source, like a garden or farm field that has been well maintained with compost over the years. Collect a third cup of sediment that is likely to be sterile, such as sand from a local hardware or landscaping company. Heat one of the soil samples in the oven to a high temperature, say 350°F, where living things are unlikely to survive.

Get four pint-sized canning jars with lids. Put each of the three samples in each of three jars. Keep the fourth jar empty of soil samples to use as our control experiment. Put a drop of bromothymol blue solution in each jar (perhaps in a small container or on a piece of glass that is laid onto the soil sample just prior to closing the jar). Note that it is critical to the experiment that each drop of bromothymol blue be of exactly the same concentration, mixed with distilled water and kept in a clean vessel and covered until use to minimize air exposure and so that no drop has been exposed to air longer than another. Seal up each jar with the lid and begin monitoring the color of the Bromothymol Blue. Record the time it takes for the drop of bromothymol blue to change color (from blue to green to yellow) in each of the jars.

Interpretation: Bromothymol blue reacts with CO_2 in air and changes color due to changes in pH. More CO_2 will result in a faster change in color. Do we see any patterns in our results that suggest that we can detect the presence of life (or at least metabolic activity) in the samples? Explain!

Our experiments will most likely have some complexities to interpret. What did we expect to happen and why? Did we see any results that we didn't expect or that seem inconsistent with the model (explanation) that we are developing? It's ok in science, and in fact desirable, to recognize when our experiments did not yield the result that we expected or the result that seems "right." What might explain our results if we do not get what we expect? Do we need to adjust the concentration of our bromothymol blue? Do we need to change how it is introduced and contained in our experiment vessel? Is our soil sample perhaps not as "living" as we might expect for a typical soil (there are more living organisms in a "living" soil sample than there are people on Earth!) We should also recognize that measuring the production of CO_2 with the bromothymol blue might not be sufficiently sensitive to see any effect even if it "seems like" we got the right answer. If so, can we figure this out from our results?

The results from the Viking landers also had some complexities, despite the fact that there were three distinctly different tests and each involved measurements more sophisticated than using bromothymol blue. Some people thought one of the tests (which introduced radioactive carbon into the samples as nutrients and then checked whether the radioactive carbon found its way into carbon dioxide in the atmosphere, suggesting metabolic conversion) showed that life had been detected. However, the other tests showed no meaningful presence of carbon in the soil to begin with, which seemed inconsistent with that result. In the end, most scientists thought, at best, the test for life was ambiguous and maybe even indicated an absence of life.

Real experimental science is about figuring what your *actual results* mean, not trying to prove some preconceived "right" answer! What do our results tell us about our experiments? Sometimes, ambiguity is the right answer! Other times, we can say things with confidence. Being able to tell the difference is a primary objective of authentic science as a practice.

Challenge 3: Identifying the Evidence for Martian life in the Story *Feedstock*.

PUZZLE 1: Below is a section of the story where the protagonist explains the evidence for life on Mars. Read the text and try to identify *exactly* what that evidence is, in other words, make a list of observations. Try to explain the evidence in your own words (for each item in your list, write why it is evidence). Can you put all the pieces together and convince one of your colleagues or classmates?

> *"Very well. It has been known for some time that many regions on Mars contain subsurface ice. Mining this ice to turn it into water is an energy-intensive process. Indeed, our present mining industries were energy-constrained until I discovered deep reserves of liquid water. If these water reserves are intelligently*

developed, Margin can look forward to many decades of lucrative work.

"Aerotek renewed my contract in order for me to assay possible liquid water deposits in your stage three mining concession, or as you call it, your Paradox project. I chose to begin by comparing known reservoirs to establish a prudent baseline case."

Hobbs, proving to be the more vocal of the corporate bookends, interrupted, "And so, after renewing your assignment, you began by wasting Margin's money, telling us what we already knew!"

"Not at all," Frank remained unruffled. "A base case provides essential calibration for any forecast. It also clearly establishes knowns, which in your case, were not so."

Culbertson coughed. "Margin was quite on top of water quality issues, Frank."

"Yes sir, but only in a qualitative sense. Margin owns exactly one fluorescence spectrometer, which hardly provides a robust method of analysis. By expanding the scope of my examination I provided the means for you to measure the scale of what now looks like a planet-wide problem." Frank paused while Dobbs looked at the ceiling and exhaled loudly, then he continued.

"I collected certain facts and mapped them. Without exception I found that in each region where Margin has discovered liquid water a distinct salinity profile exists. Whereas your shallow exploratory wells tapped fresh waters, your later and deeper wells found progressively higher salinities. Your own corrosion maintenance reports bear this out."

Culbertson shifted his overfed frame, creaking his chair. "Frank, we all know the history of water on Mars. There is no need to give us a miner's introductory course."

"Yes, we all know that the water essential for any expansion of our operations on Mars originally came from glaciation. Our best estimates put last-stage glaciation from 50,000 to 100,000 years ago. But gentlemen, glaciers consist of frozen fresh water. Where did the salt come from?"

"Oh come now," Culbertson frowned, "surely no one disputes that salt came from the bedrock itself as water flowed through it? Do you challenge that?"

"In your stage one mining area my work clearly shows that, in fact, salts dissolved out of the host rock. But in all other cases, and in the rest of Margin's concessions, no suitable rock formation exists to provide the salts. Further, it is only in the remaining cases that we find clathrates."

"What?" Dobbs shook off his bored stupor at the unfamiliar word.

"Gas hydrates. Think of clathrates as ice cages. Microscopically, they resemble linked geodesic spheres, each just big enough to contain one molecule of methane gas. We find them at the extreme updip ends of the water reservoirs, closest to the cold surface of the planet."

Culbertson addressed an aside to Dobbs: "We know all about these gas-ice hydrate deposits. They've never proven to be a problem for our operations."

"And yet," Frank said, "no one before me attempted to analyze the gas that they contain. Isotopically, the methane gas trapped

in these small ice cages is identical to Archaebacteria-sourced gas on earth."

"Who cares?" Hobbs was back on point.

"We should all care, Mr. Hobbs, because all lines of evidence converge. Take the organic carbon values of the seal rocks encasing your aquifers, for example. Some measurements exceed 20% carbon by weight. Or look at the obvious link between bacterial generation of methane and bicarbonate production. This ties to the increased alkalinity due to carbon dioxide being turned into bicarbonate at the bottom of your reservoirs."

Spoiler Alert: Here are our initial thoughts on which passages are actually evidence, which are conclusions, which are reasoning or arguments connecting the evidence to the conclusion, and which are related assertions or background information from prior studies.

Interpretation of Selected Text from the First Passage from *Feedstock*	
Text from the Story	**Interpretation of whether the text is Evidence, Conclusion, Reasoning, or Something Else**
many regions on Mars contain subsurface ice	An assertion based on prior studies
what now looks like a planet-wide problem	A conclusion, but the exact conclusion and any evidence for it has not yet been cited in the story.
where Margin has discovered liquid water a distinct salinity profile exists	A conclusion based on his sampling and testing, although the exact nature of the conclusion is not yet stated.
I collected certain facts and mapped them.	Evidence for the 'Distinct salinity profile", although the exact nature of the measurements is not revealed.

later and deeper wells found progressively higher salinities	More details of the conclusion are revealed—higher salinities at depth.
our own corrosion maintenance reports bear this out	Further evidence for higher salinities at depth, implying that corrosion records show increased salinities at depth.
the water essential for any expansion of our operations on Mars originally came from glaciation	Assertion based on prior studies
But gentlemen, glaciers consist of frozen fresh water. Where did the salt come from	Argument/reasoning relating to the increased salinities at depth
surely no one disputes that salt came from the bedrock itself as water flowed through it	Statement of the explanatory model accepted prior to the new evidence presented in the story
no suitable rock formation exists to provide the salts	Evidence (observation of no suitable rock) combined with reasoning (that the absence of a salt source means that the salt can't come from the bedrock)
it is only in the remaining cases that we find clathrates.	Evidence (observational data) for a conclusion that has not yet been identified in the story
Gas hydrates. Think of clathrates as ice cages. Microscopically, they resemble linked geodesic spheres	Ancillary background information based on prior studies
We find them at the extreme updip ends of the water reservoirs, closest to the cold surface of the planet.	Evidence, adding field situational evidence to the chemical evidence of simply noting the presence of clathrates.
Isotopically, the methane gas trapped in these small ice cages is identical to	Evidence, measurement of the isotopic character of the methane in clathrates, and the start of

Archaebacteria-sourced gas on earth	argument/reasoning tying that observation to a possible conclusion (living things may have produced it)
Take the organic carbon values of the seal rocks encasing your aquifers, for example. Some measurements exceed 20% carbon by weight.	Evidence, observation of high C contents
look at the obvious link between bacterial generation of methane and bicarbonate production	Reference to prior knowledge as part of making an argument/reasoning to connect observed bicarbonate to living things.
ties to the increased alkalinity due to carbon dioxide being turned into bicarbonate at the bottom of your reservoirs	Evidence, observation of bicarbonate that has already been linked to living things through argument/reasoning in the preceding text.

PUZZLE 2: Below is the paragraph where the protagonist summarizes the evidence. How does the summary compare with the list you constructed for evidence for life? Does it offer new evidence that wasn't cited before? Is there evidence cited before that isn't repeated here? Does the summary help you see the evidence more clearly or in a new light?

> *"Yes sir. By alien life forms comparable to Archaebacteria, living in subsurface communities in fresh water halos above the alkaline waters. I've spent the last two years testing water samples across your concession areas and they clearly show the alien biota as assemblages of flimsy cells and filaments no different from their counterparts on earth. The microbes appear to be eating the organic matter in the high-carbon shales which seal your aquifers. They excrete both carbon dioxide gas, which drives the bicarbonate formation, and methane gas, which ascends to take up residence in the ice cages."*

Spoiler Alert: Here are our initial thoughts on which passages are evidence, which are conclusions, which are reasoning or arguments connecting the evidence to the conclusion, and which are related assertions or background information from prior studies in the second "summary" passage from *Feedstock.*

Interpretation of Selected Text from the Second Passage from *Feedstock*	
Text from the Story	**Interpretation of whether the text is Evidence, Conclusion, Reasoning, or Something Else**
By alien life forms comparable to Archaebacteria, living in subsurface communities in fresh water halos above the alkaline waters	Primary conclusion, along with ancillary information that will be tied to the evidence and reasoning.
I've spent the last two years testing water samples across your concession areas and they clearly show the alien biota as assemblages of flimsy cells and filaments no different from their counterparts on earth	Evidence, observation of cells and filaments. This observation was not included in the text of the first passage.
The microbes appear to be eating the organic matter in the high-carbon shales which seal your aquifers	Reasoning/conclusion tying the observed field association with high-organic rock, cited in the first passage, with the primary conclusion of the presence of microbes.
They excrete both carbon dioxide gas, which drives the bicarbonate formation, and methane gas, which ascends to take up residence in the ice cages	Stated as a conclusion, but indirectly referring to the primary evidence cited in the first passage—that is, the presence of bicarbonate and the methane gas whose isotopic character suggested a biological origin, evidence which is not repeated here.

Reflection on Learning

The goal of this book is to practice investigation itself, not to focus on the information and ideas you understood better through investigation. If traditional learning outcomes are not the goal, then what have you learned? Taking some time to consider what you gained from your investigation, reflecting on the practice of science *metacognitively*, can help you hold onto what you have gained. For example: What questions did you ask? What observations and interpretations did you make? Did you *do* modeling, rather than just learning or applying a model?

Real investigation is fluid, creative, and defies easy categorization. Even so, it might be helpful to consider the following categories of investigation offered by the Next Generation Science Standard's science and engineering practices (NGSS Lead States. 2013. Next Generation Science Standards: National Academies Press).

- Asking questions and defining problems
- Developing and using models
- Planning and carrying out investigations
- Analyzing and interpreting data
- Using mathematics and computational thinking
- Constructing explanations and designing solutions
- Engaging in argument from evidence
- Obtaining, evaluating and communicating information

Notice that we have not numbered these practices. They are not steps in a sequential process, but rather components of an integrated and iterative process. Also recognize that this list of practices is **a** way to organize and understand the investigative practice of science, not **the** way. As you go through this reflection, be flexible in thinking about what aspects of your own investigation might be practices of science, because these practices can show up in unexpected ways. Remember, don't think so much about

what you learned or understood, but rather think about what you initiated, did, practiced or experienced.

On a separate sheet of paper, jot down some things that have strengthened your ability to pursue investigative thinking. Give examples of how you engaged in particular, specific practices.

Exploring with you the experience that comes from this particular Learn by Example might provide some guidance as you think about your own practices of science. What practices of science might someone working through these challenges experience?

The practices of science associated with this Learn by Example in the Table of Contents include discussion, experimentation, and interpretation. These three are related in a fairly straightforward manner to the three Challenges presented.

Challenge 1, Discussion of Mining Morality, engages in discussion of a complex topic of relevance in the modern economic and political world. Preparation for this discussion, and engagement in it, might involve practices such as asking questions and defining problems, obtaining, evaluating and communicating information, constructing explanations and designing solutions, and arguing from evidence.

Challenge 2, Experimental Investigation of Biological Activity, offers a general concept that would need to be developed in a similar fashion to the experimental design for the Learn by Example for the story "Breaking the Ice." If done properly, such open-ended experimental investigations engage the investigator with all of the practices of science.

Challenge 3, Identifying the Evidence for Martian Life engages with interpretation at many levels, including interpretation of the text so as to identify what is evidence and what is conclusion (not at all a trivial task!). It also engages the investigator with interpretation of the observational evidence itself, which the story does not typically explain.

Most significantly, the Challenges in this last chapter are substantially open-ended, inviting you, the reader, toward further exploration and investigations of your own. The investigations presented in this chapter are only "starter" investigations, a launching place for the next step in discovery.

In the end, every investigation is a launching place for the next step. As Isaac Newton famously said, "we stand on the shoulders of giants. No investigation is ever the starting place or the ending place. Every investigation is somewhere in the middle, a place for us to start as we engage with the practices of science that have allowed people over thousands of years to begin to understand our universe.

The most interesting learning outcome in science is not what we have learned so far. The most interesting learning outcome is *how we learn.* We hope that this book has given you deeper insight into how that happens and greater experience with the practices of science that have moved us forward in investigative thinking to the place where we are today.

About the Editors

Russ and Mary Colson, the editors of this book and authors of the investigative lessons, are a husband-and-wife team who met in geology graduate school at the University of Tennessee. Mary earned a Master of Science degree in sedimentology, taught 8th grade earth science for thirty years, was named Medtronic Minnesota Middle School Science Teacher of the Year in 2008, and was one of the writers of the Next Generation Science Standards which has influenced teaching across the country for more than a decade. Russ earned a PhD in experimental planetary science, did research at the Johnson Space Center and Washington University in St Louis, and taught college geology at Minnesota State University Moorhead for thirty years where he established the program in Earth Science Teaching and was honored as national professor of the year in 2010. They have worked as a team on many projects over the years, including coauthoring the book *Learning to Read the Earth and Sky* (published by the National Science Teachers Association Press), publishing peer-reviewed research in the widely-different fields of experimental geochemistry and dinosaurs, and offering science teacher workshops on investigative learning in their home state of Minnesota, which they continue to do in retirement. They also continue to publish an online magazine, *Issues in Earth Science*, on which this book is based, which features earth-science related YA fiction stories with accompanying classroom activities. In retirement, they garden on their rural property in NW Minnesota, Russ writes science fiction novels and short stories which you can check out on his website at russcolson.com, and Mary quilts and is learning to play the violin!